NCC Blackwell
MANCHESTER·OXFORD

British Library Cataloguing in Publication Data
Skidmore, S.
  SSADM.
  1. Management. Information systems. Application of
  computer systems. Structured systems analysis & systems
  design
  I. Title   II. Farmer, R.   III. Mills, G.
  658.40380285421

  ISBN 0–85012–796–3

Published for NCC Publications by NCC Blackwell Limited.

Editorial Office, The National Computing Centre Limited,
Oxford Road, Manchester M1 7ED, England.
NCC Blackwell Limited, 108 Cowley Road, Oxford OX4 1JF, England.

Typeset in England by Bookworm Typesetting, Manchester
and printed by Hobbs the Printers Ltd of Southampton

ISBN 0–85012–796–3

Preface

The Structured Systems Analysis and Design Method (SSADM) is a set of procedural, technical and documentation standards for systems development. It is owned by the Central Computer and Telecommunications Agency (CCTA) which is currently part of HM Treasury and the centre of information system policy and procurement in government. CCTA has adopted an 'open' strategy in which SSADM is non-proprietary and is publicly available.

SSADM has a rigorous release strategy with the broad content of the next version published about nine months in advance of its actual release. This book examines Version 4, which was launched by the CCTA in May 1990.

SSADM strategy is determined by a Design Authority Board comprising members of the CCTA, the Computing Services Association (CSA), the British Computer Society (BCS) and the National Computing Centre (NCC). The official reference manual is published by NCC-Blackwell.

The Systems Analysis Examination Board (SAEB) of the BCS controls examination standards in SSADM and awards certificates of proficiency to successful candidates.

SCOPE OF THIS BOOK

This book considers the construction and use of the fundamental models of SSADM. Each technique is examined in depth, giving detailed guidelines on notation, construction, presentation and documentation. The progressive development of each model is illustrated using detailed examples from a selection of applications. Chapter 12 is a brief introduction to the framework of application, showing where each model and technique appears in SSADM and illustrating the relationships between them.

This emphasis on the products of models of SSADM has been adopted for three main reasons. The first is that the management framework is explored in detail in complementary NCC publications (see References) and these will be of interest to the practitioner. Secondly, the techniques can be used independently of the framework and this may be of particular significance to teachers and lecturers in further and higher education who wish to use the **notation** of SSADM but not its prescription. SSADM employs methods which are also fundamental to other structured methodologies and so there is not an inextricable link between the modelling methods and their framework of application.

Finally, SSADM Version 4 is a product oriented methodology with progress determined and monitored through the production of models defined to a particular standard and quality.

The NCC-Blackwell Reference Manual is particularly comprehensive and it is not the purpose of this book to provide a "cut-down" summary of it. Figure 1 (reproduced from the Reference Manual) shows the three main components of SSADM:

- the Dictionary defining 'what' should be produced;
- the Structural Model describing 'when' to produce it;
- the Procedural Chapters explaining 'how' to do it.

This book concentrates on the 'how' of SSADM for most of the products defined in the Dictionary. The index of Product Descriptions of the Reference Manual is reproduced at the end of this Preface (Figure 3) together with page references for this book. This allows readers to assess the product scope of the book. The 'when' perspective of the Structural Model is confined to Chapters 1 and 12. Cross references between the text and the Structural Model are included in the Structural Model Diagrams of Chapter 12.

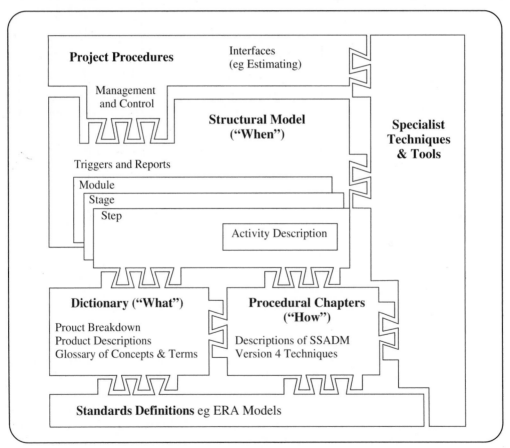

Figure 1   The three main components of SSADM

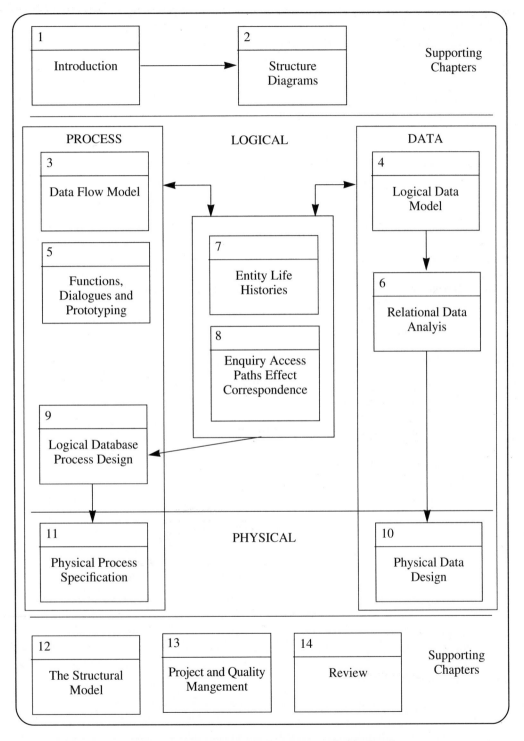

Figure 2   Content & Structure of this Book

CONTENT OF THE BOOK

A diagram showing the content and structure of this book is shown in Figure 2. A brief introduction to each of the principal chapters is given below:

Chapter 1   Introduction

Chapter 2   Structure Diagrams

The SSADM Structure Diagram is used in several different techniques. Chapter 2 describes the notation in general to avoid unnecessary repetition in subsequent chapters. Top-down and bottom-up approaches to deriving the diagram are both examined because neither approach is always superior in all circumstances.

Chapter 3   Data Flow Model

Data Flow Diagrams (DFD) show the passage of data through the system. They focus on the processes that transform incoming data flows into outgoing ones. DFDs are used in most structured methodologies and indeed were the principal model of the two books that effectively launched structured analysis (*deMarco 1979, Gane and Sarson, 1979*). Chapter 3 examines the development of a DFD from two perspectives.

– Bottom-up. This traces the construction of a DFD from its use as a description of the current system (so-called current physical) through its logical equivalent (current logical) to encompassing agreed user requirements (required logical).

– Top-down. General principles of DFD development are given starting with a consideration of the inputs and outputs of the system. This is used to define a one process DFD (or context diagram) which is progressively developed and decomposed.

Data Flow Diagrams are supported by various documentation which contribute to the completion of the Data Flow Model.

Chapter 4   Logical Data Model

The Logical Data Structure (LDS) is a static representation of the data used in the system. It shows how the data is logically grouped and how these logical groupings are related to each other. Similar models are found in most structured methodologies. Chapter 4 introduces the notation and construction of these models together with a brief examination of the documentation used to support the LDS in the Logical Data Model.

Chapter 5   Functions, Dialogues and Prototyping

The processes of the DFD describe the main functional requirements of the system. However these processes are not constructed within a business or user context and hence do not necessarily reflect how the user perceives the functionality of the system. Consequently SSADM has introduced the definition of functions as a super-technique for describing sets of system processing which the user wishes to carry out at the same time to support a certain business activity.

The interaction between the user and the function is modelled with Dialogue Structures and can be validated and explored with prototype construction and experimentation. This chapter looks at the construction of functions, dialogue models and prototype control and documentation.

Chapter 6   Relational Data Analysis

The logical groupings of data described in the Logical Data Model are most useful when they are well normalised. Chapter 6 examines the construction of data structures which are unambiguous, free from redundant duplication, flexible and easy to maintain. This is achieved through Relational Data Analysis (RDA). Two detailed examples are given.

The first is a complex data structure which is normalised using a formal consideration of the functional dependencies between data items. The second example demonstrates the normalisation of a relatively simple data set using a cook-book approach. A set of rules for converting the normalised data groups into a Logical Data Model is also provided.

Chapter 7   Entity-Event modelling: Entity Life Histories

Entity-event modelling provides a third view of system requirements. It is concerned with identifying events, the sequence in which these events occur and the effect these have on the data held in the system. It is often represented as a "time perspective" of the system and is used to amend or extend the process and data models.

Two modelling techniques are used in entity-event modelling. Chapter 7 describes the Entity Life History (ELH) which is used to investigate and document the life of each entity contained in the Logical Data Model. It graphically illustrates how occurrences of each entity are created, modified and deleted.

Chapter 8   Enquiry Access Paths and Effect Correspondence Diagrams

Each update and enquiry function requires an access path through the Logical Data Structure in order to locate the required data. Enquiries are modelled using Enquiry Access Paths and update functions are defined in Effect Correspondence Diagrams (ECD). The ECD is the second modelling technique of entity-event modelling. These are developed from the Entity Life Histories and highlight how the effects on entities are related to each other.

Chapter 8 illustrates the development of Enquiry Access Paths and Effect Correspondence Diagrams for the Entity Life Histories constructed in Chapter 7.

Chapter 9   Logical Database Process Design

Logical Database Process Design produces Enquiry Process Models (EPM) and Update Process Models (UPM) which provide a detailed process specification. They are largely developed from the Enquiry Access Paths and Effect Correspondence Diagrams described in the previous chapter.

This chapter looks at the development of the two models as well as examining general issues of process design which must be taken into account in their construction.

Chapter 10   Physical Data Design

Physical Data Design is concerned with translating the data groups of the Logical Data Structure into a physical file and data base design for a specific implementation environment. Chapter 10 examines a general strategy for moving from the logical to the physical data model and encourages the development of a design strategy that formally documents local experiences and knowledge.

Chapter 11   Physical Process Specification

Physical Process Design is concerned with the conversion of logical processes and functions into program specifications, physical input/output formats and physical dialogues. The detail of the conversion obviously depends upon the target hardware and software environment and so a general strategy can only be described. This chapter examines many of the issues involved and again emphasises the importance of developing a formal installation design strategy that takes into account local knowledge and experience.

Chapter 12   The Structural Model

This chapter provides a brief introduction to the modules which make up the Structural Model of SSADM. Each module is described with reference to its objectives, steps, products and primary techniques. Most of the products and techniques listed in the module descriptions can be directly cross-referenced back to the contents of Chapters 3-11.

Chapter 13   Project and Quality Management

This chapter examines selected issues of project management and quality control. It is not a detailed examination of either of these areas but rather a summary of the project and quality controls that will surround the products described in this book.

Chapter 14   Review

CONTEXT OF THE BOOK

This book concentrates on the fundamental modelling methods of SSADM, not the fundamental skills of analysis and design. It must be recognised that SSADM demands an underpinning of good principles and practice in analysis and design. This is explicitly recognised in the Certificate of Proficiency in SSADM, where an oral examination is designed to:

> "assess whether a candidate has the skills, knowledge and personal qualities to enable him or her to make an effective contribution to systems analysis and design generally".

This particular text builds upon the foundations laid in *Introducing Systems Analysis* and *Introducing Systems Design,* Skidmore and Wroe (1988,89).

Finally, this book does not attempt to justify the use of SSADM or compare it to other methodologies. It assumes that the decision to implement, teach or research SSADM has already been taken.

Figure 3   Cross-reference to SSADM products

Contents

1 Introduction

1.1 SSADM

The Structured Systems Analysis and Design Method (SSADM) is a set of procedural, technical and documentation standards for systems development. It is owned by the Central Computer and Telecommunications Agency (CCTA) which is currently part of HM Treasury and the centre of information system policy and procurement in government. CCTA has adopted an 'open' strategy in which SSADM is non-proprietary and is publicly available.

SSADM has a rigorous release strategy with the broad content of the next version published about nine months in advance of its actual release. This book examines Version 4 which was launched by the CCTA in May 1990.

SSADM strategy is determined by a Design Authority Board comprising members of the CCTA, the Computing Services Association (CSA), the British Computer Society (BCS) and the National Computing Centre (NCC). The official reference manual is published by NCC Blackwell.

The Systems Analysis Examination Board (SAEB) of the BCS controls examination standards in SSADM and awards certificates of proficiency to successful candidates.

At the time of writing, British Standards Institute (BSI) committee IST/5 (software development and systems documentation) have prepared a draft standard for public comment. The British Computer Society is represented on the committee and has supervised the work. This draft is in four parts numbered 91/64274, 64275, 64277, 64278 and these cover the overall framework, the modelling concepts, basic product descriptions, and modules and stages respectively.

1.2 SSADM WITHIN THE DEVELOPMENT LIFE CYCLE

SSADM Version 4 comprises five core modules (see Figure 1.1) which will be supplemented over time by Study Guides. The place of these modules within the whole life cycle is shown in Figure 1.2. Strategic planning is still outside the scope of core SSADM, although some of the techniques described in this book (notably the Logical Data Model) can make significant contributions to strategic decisions.

The extent of SSADM's contribution to Physical Design depends upon the implementation

1

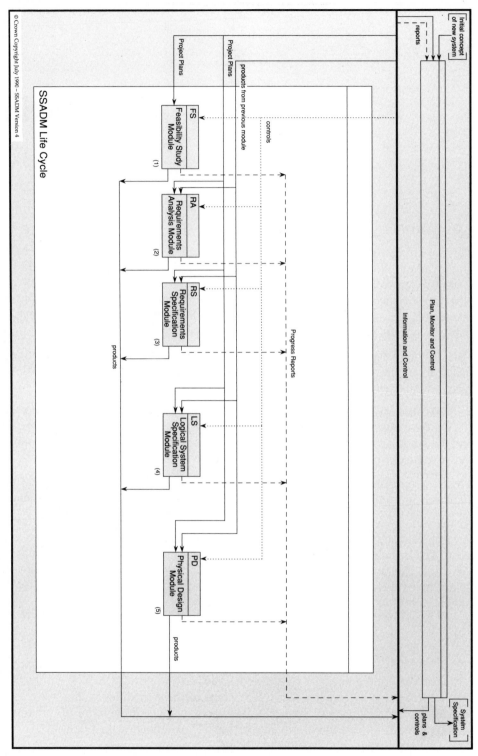

Figure 1.1   SSADM core method

language. If a third generation language such as COBOL is being used, then SSADM ends at program design. However, if a fourth generation language is being employed, then design ends with an application which is fully defined in an appropriate non-procedural language.

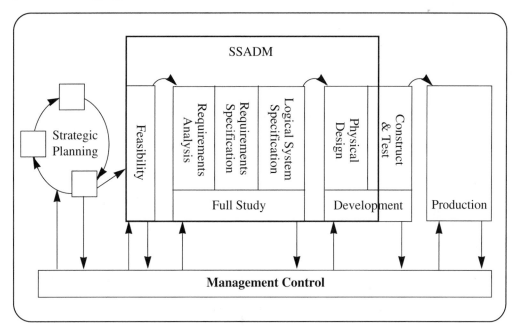

Figure 1.2   Position of SSADM in the life cycle

1.3   THE PRODUCT PERSPECTIVE

Each Module of SSADM is broken down into Stages and Steps. All SSADM activities, whether they are Modules, Stages or Steps, are considered as 'black boxes' in that they will produce predictable outputs from given inputs. The inputs and outputs consist of the defined **products** of SSADM. This product oriented approach has a number of advantages:

- Project planning concentrates on identifying the required deliverables to fulfil the project objectives. This leads to defining the products and the steps and tasks necessary to make these products to a defined quality criteria. Hence, project progress is tied to deliverables rather than to the passage of time or activities. Developers are aware of what products are required, and to what standard.

- The development task can be split between suppliers or internal teams. If the products of a certain activity are accepted, then there is no need for the supplier of the succeeding activity to know anything about the internal construction or process of its predecessor. Programmers will recognise the principles of modular programming here. A called module receives data from the calling module but does not have to know anything about the internal construction of that module to decide what to do with the received data.

- Specialised teams can be developed with responsibility for certain activities across projects. A particular example is Physical Design, where detailed knowledge of the implementation environment can be built up in a small specialised team.

- Project estimation is based on tangible deliverables rather than notional percentage time estimates.

- Established quality criteria can be applied for each product. The Product Descriptions defined in the Reference Manual have defined quality criteria.

1.4   THE PRODUCT BREAKDOWN STRUCTURE (PBS)

This section introduces the products of SSADM and shows how they fit into a standard model of system development, control and quality assurance. This model consists of products arranged within a hierarchy, called a Product Breakdown Structure (PBS).

Within the context of this book the PBS is used to:

- provide the context of SSADM. It lists products which are not necessarily prescribed within SSADM but which are needed for information systems development and delivery;

- introduce the products of SSADM and defines where they are considered in the book;

- define the development life cycle of SSADM, briefly describing the content and structure of the five modules of core SSADM.

The PBS described here is only an example and can obviously be tailored to local requirements. However, it serves to illustrate the main points.

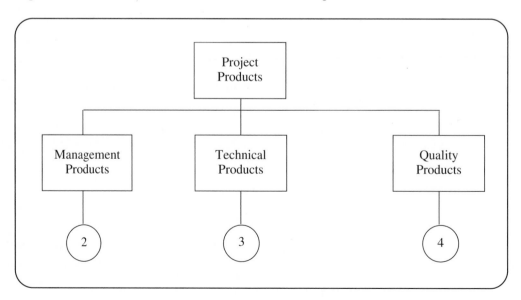

Figure 1.3   Top level Product Breakdown Structure

1.4.1 Project Products

The top level of the PBS consists of three complementary product categories (see Figure 1.3).

- *Management products*. These are used in the planning and control of the project. Typical products include:
  - Installation development standards;
  - Project Initiation Document;
  - Project Plans;
  - Project Board minutes;

Quality Criteria:

For all:

1	Is the variant identifier properly completed?
2	Are notational conventions correctly applied?
3	Is the boundary of the system clear?
4	Are all user-perceived functional areas represented?
5	Are meaningful names used for processes and data stores?
6	Are all identifiers unique?
7	Do external entity names accurately reflect the environment outside the system?
8	Does the diagram accurately reflect the physical, logical or required system in terms of external entities, data stores and data flows?
9	Does the diagram avoid giving an inappropriate level of detail, such as sequencing or detailed processing logic?

For Logical Data Flow Diagram:

10	Are all physical aspects of the current system removed, unless they are constraints on the requirement?
11	Are any enquiries remaining after logicalisation major ones?

For Required System Data Flow Diagram:

12	Are all the facilities defined by the Selected Business System Option, and nothing else, modelled in the Required System Data Flow Diagrams?

Figure 1.4   Example quality criteria for Data Flow Diagrams

– Checkpoint and project status reports;

– Project Evaluation Review;

– Post Implementation Review;

The project plans will include activity descriptions, networks and product flow diagrams. SSADM does not address most of the issues of project planning and control directly. It assumes that these are dealt with in a complementary project control method such as PRINCE (see References). Certain aspects of project planning are discussed in Chapter 13 but it is not a major concern of this book. Indeed, one of the significant changes introduced in Version 4 is the separation of technical SSADM activities from those of project management.

– *Technical products.* These contain the major products of the development process. The production of these models is the main focus of this book and hence this category is considered in more detail below.

– *Quality products.* These document the quality of systems development. Typical products include:

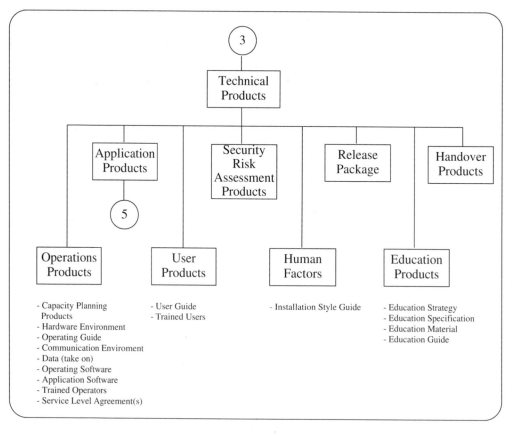

Figure 1.5   Technical Products of SSADM

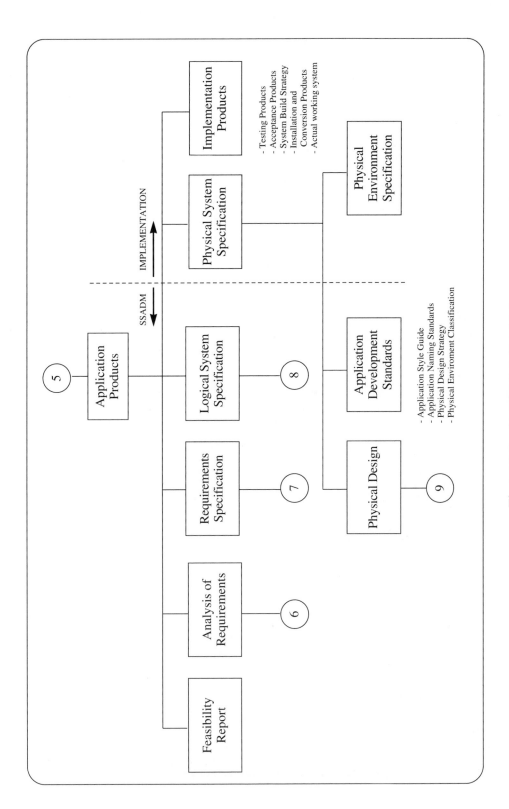

Figure 1.6   Application Products

- Product descriptions. Each Product Description defines quality criteria against which each individual product must be assessed (see Figure 1.4 for an example).

- Quality Review Agendas and meeting records.

- Project problem records. These will be used to feedback into individual projects, project control procedures and quality assurance.

SSADM does not directly provide procedures for quality management, assurance and control. It assumes that complementary quality programmes and procedures are in place. These may be defined internally or, preferably, may be agreed to BSI or ISO standards. SSADM's product quality criteria should promote better quality management and hence facilitate adoption of BS 5750 or ISO 9001. Some issues of quality are discussed in Chapter 13.

1.4.2   Technical products

The Technical Products of SSADM fall into eight categories (see Figure 1.5). The Application Products are our main concern here, although the Installation Style Guide is discussed in Chapter 5 and the constraints and policies embodied in Operations Products are encountered in Chapters 10 and 11.

The Application Products (see Figure 1.6) are essentially the analysis, design and implementation products. This is the point where SSADM specific products fit into the Product Breakdown Structure. The aim of this book is to demonstrate the construction of some of these products.

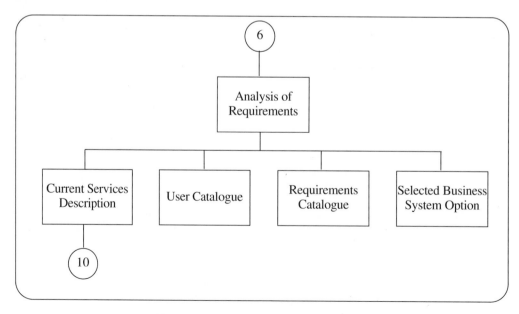

Figure 1.7   Requirements Analysis

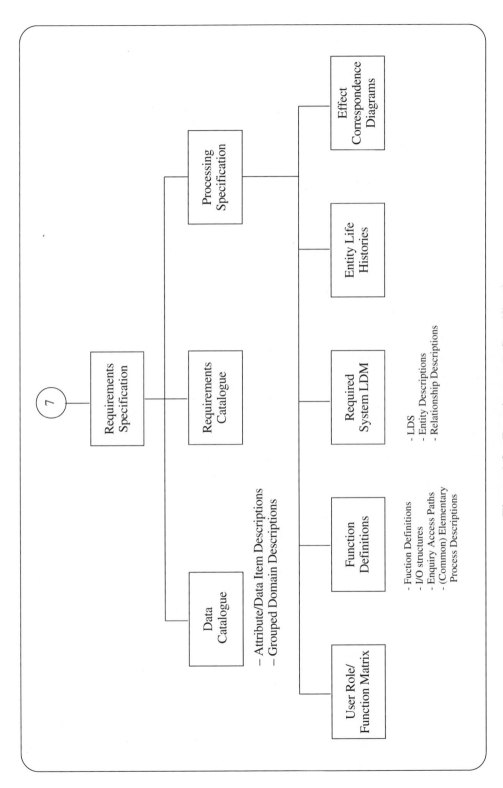

Figure 1.8   Requirements Specification

1.4.3   Application Products

Feasibility Report

This summarises whether user's business requirements can be economically, technically and operationally satisfied. It is the main product of the Feasibility Study Module and its recommendations will determine whether full-scale analysis and design will commence. Most of the products of this Module are first attempts or cut-down versions of models used in Requirements Analysis and Specification.

Requirements Analysis

Figure 1.7 shows the module products for the Requirements Analysis Module.

The *Current Services Description* (See Figure 1.13) is a logical view of the existing information processing systems. The primary products are the Current Environment Logical Data Model (LDM) and the Logical Data Flow Model (DFM). These two perspectives of the current system are associated through a logical datastore/entity cross-reference and both are underpinned by a shared Data Catalogue. These models may be based on preliminary work undertaken in the Feasibility Study.

The *User Catalogue* summarises the activities and responsibilities of users and potential users of the system. The *Requirements Catalogue* lists the problems of the current information system and the requirements of its successor. Both of these catalogues are active documents which are updated as development progresses.

The business implications of different system boundaries are explored in the Requirements Analysis Module. The selected option is then documented and presented to user management.

Requirements Specification

The products of the Requirements Specification Module are summarised in Figure 1.8. In this module, the developer examines the processing specification in depth and begins to consider how that processing will be presented to the user.

The LDM is extended to cover the scope of the proposed or required system. The Data Flow Model (DFM) is used as a basis for Function Definitions. These are units of processing defined from a user perspective and hence their definition requires the active participation of users.

It is also in this module that the third perspective of the system (time and events) is developed. The resultant products are Entity Life Histories and Effect Correspondence Diagrams.

Logical System Specification

The Logical System Specification Module begins with the definition of alternative technical system options that satisfy the Requirements Specification. One is selected and documented fully in a Technical Environment Description (see Figure 1.9).

The Logical Design (see Figure 1.10) gives a detailed logical definition of the required system. The data perspective is provided by the Required System LDM and the process view by a set of logical process models based on the functions and event-entity models of the requirements specification. The dialogue structures are also defined more fully, building on the I/O structures and User Role/Function Matrix of the previous module.

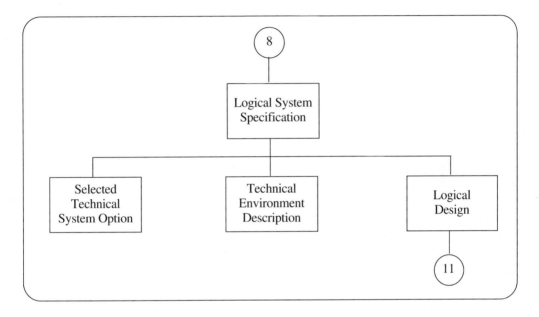

Figure 1.9   Logical Systems Specification

Physical Design

The Physical Design Module takes the products of logical design and produces specifications for the target hardware and software environment. The products of this Module are summarised in Figure 1.11. The detailed data and processing specifications will be supported by other specification and installation standards developed to reflect local requirements.

1.5   THE STRUCTURAL MODEL

The Structural Model formally defines the activities within SSADM and the relationship between those activities. These relationships are represented by product flows on a Structural Model Diagram (see Figure 1.12). The Structural Model consists of a number of hierarchical diagrams representing Modules, Stages and Steps; Figure 1.12 is a typical Step definition. This type of definition is known as an Activity Description, and defines the tasks to produce the required products of that activity. Activity Descriptions are defined in logical sequence in an Activity Network to describe the plan for the complete project.

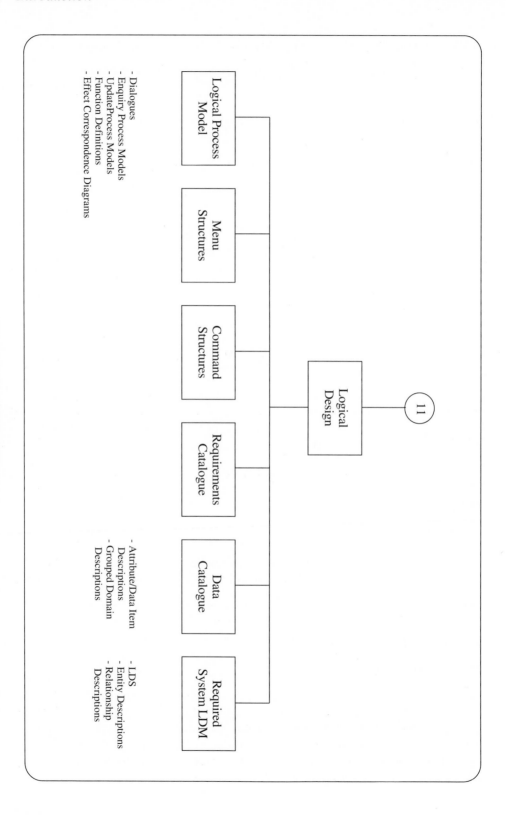

Figure 1.10   Logical Design

In the context of the content of Activity Descriptions, this book focuses on the products, techniques and (to a lesser extent) activities.

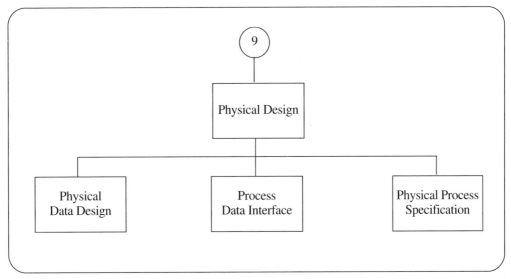

Figure 1.11　Physical Design

The Structural Model is supported by Structural Model Diagrams (see Figure 1.12). The information highway defined on this diagram is the communications route for all product and control flows between SSADM Modules. The SSADM practitioner is concerned with those activities below the information highway. These processes are decomposed into subordinate processes showing the flow relationships, but not the internal process detail. Three types of flows are distinguished:

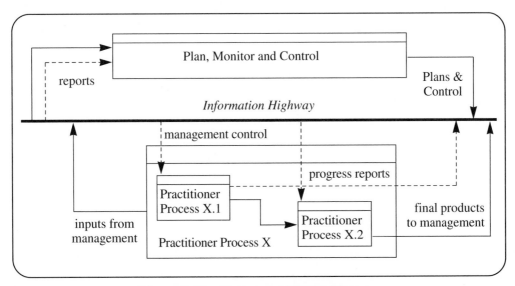

Figure 1.12　Structural Model Diagram

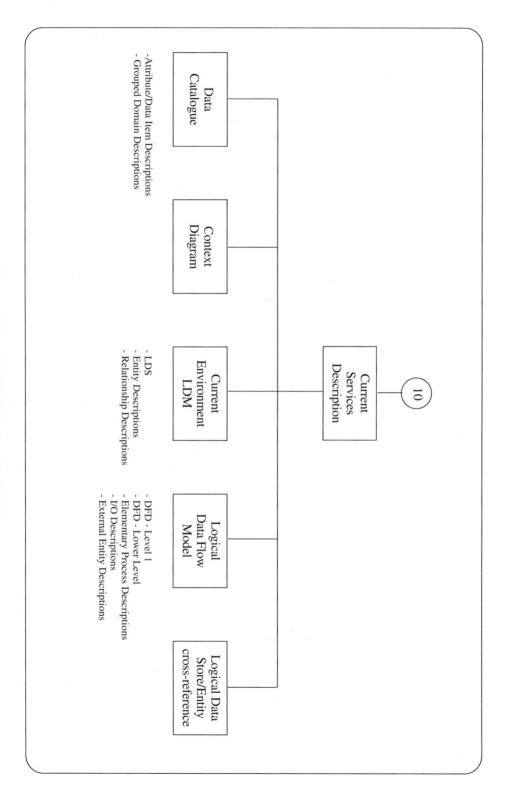

Figure 1.13   Current Services Description

————————————> Product or activity flow

· · · · · · · · · · · · · · · ·>Control or management authorisation flow

— — — — — — — — — — — > Progress reports. These are usually omitted from the lowest level diagrams to aid clarity.

SSADM products defined in Product Descriptions (NCC, 1990) are shown with initial capitals. These products are always defined at the highest level on the Product Breakdown Structure. So, for example, a flow described as Current Environment LDM would consist of the products defined on the PBS (see Figure 1.13), in this case a Current Environment LDS, Entity Descriptions and Relationship Descriptions.

The Structural Model Diagram uses shading to highlight information and the left to right ordering represents time progression, but not time intervals. Parallel steps are boxed together.

The Structural Model of SSADM is described informally in Chapter 12, supported by annotated Structure Model Diagrams. The overall Structural Model Diagram for the five modules of SSADM is repeated in Figure 1.14.

1.6   SSADM MODELLING PERSPECTIVES

SSADM views the system from three perspectives:

- *Functions*

    Function Definitions summarise the user's view of system processing. The functional perspective is initially created from the Data Flow Model (DFM). However, the DFM does not show how procedures apply to different users or how procedures react to different events. Consequently, dialogue design is used to tackle issues of human-computer interaction and entity-event modelling is employed to investigate system processing. Function Definition coordinates these activities and is a significant introduction into Version 4.

- *Events*

    Events are business activities which trigger processes to update system data. An event is not the process itself but the stimulus that causes a process to be invoked, for example . . . receipt of application form, all cargo cleared from a ship, seven years since invoice date etc . . . Events are modelled through entity-event modelling which has two major techniques: Entity Life Histories (ELHs) and Effect Correspondence Diagrams (ECDs). The latter is newly introduced into Version 4.

- *Data*

    The data perspective is represented by Logical Data Models (LDM) and Relational Data Analysis (RDA). This is a well established viewpoint and provides a particularly stable view of the system. Version 4 gives a more flexible approach to the complementary use of these two techniques.

These three perspectives are complementary. Knowledge gained in one area must feed back into the other two. Definition of requirements occurs incrementally and is recorded in the Requirements Catalogue, which is a central document throughout the project.

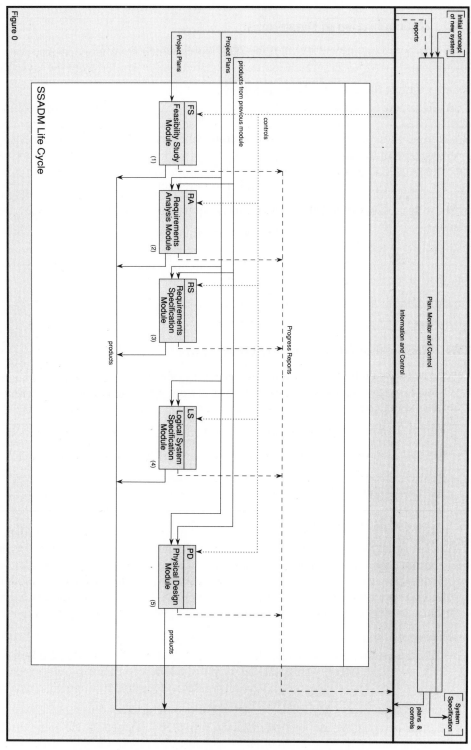

© Crown Copyright July 1990 – SSADM Version 4

Figure 0

SSADM Life Cycle

Figure 1.14   Overall Structural Model Diagram

1.7   LOGICAL MODELLING

A particular problem of systems development is the transition from the analysis of a current system to the design of its replacement. This is true whether the current system is a manual system, a computer system or some combination of the two.

The structured techniques appeared at a time when there was growing dissatisfaction with an approach to analysis that placed considerable emphasis on fact collection rather than system development. In this conventional method there was no clear way of making the progression from analysis to design. The new system often inherited unnecessary operational characteristics of its predecessor, and so the application changed in technology but not in scope and operation.

At the same time, there were considerable changes in software technology. Fourth Generation Languages (4GLs) and Data Base Management Systems (DBMS) provided new opportunities and difficulties. New hardware possibilities also emerged. Powerful PCs, PC-Networks and PC links to mainframes changed the face of computing. Applications were no longer restricted to COBOL systems on large mainframes.

Many organisations also took the opportunity to re-assess their approach to information system development. Most of the operational transaction-based systems were already computerised, or could be implemented with proprietary software packages. Management began to look to computerisation to provide a competitive business edge. It was increasingly recognised that the enterprise's data was a significant resource that should be managed. It might also provide new business opportunities, exploiting market-places which could not be accessed through conventional technology.

Thus a set of models were required that were:
- graphical, based on the principle that a picture is worth a thousand words;
- not constrained by current organisation structures, physical arrangements and restrictions due to machine type and capacity;
- business system and solution orientated. Conventional flowcharts tended to trap developers into premature computer solutions and were inaccessible to system owners and developers.

Logical models were developed to meet these requirements. These showed the information system stripped of administrative arrangements and physical trappings. They focussed on why an activity takes place and what it does, not on who, where and how it is done.They provide a firm basis for design because they concentrate on the fundamental requirements of the system, uninhibited by constraints of hardware and software.

Physical models are used in Stages 1 and 6 of SSADM. Step 130 of Stage 1 concerns the production of Data Flow Diagrams of the current physical system. These focus on how the process is done, where and by whom.

Stage 6 is concerned with the definition of files and programs. These definitions will reflect the physical constraints of the target hardware and software environment.

However, in between, SSADM – in common with other methodologies – is overwhelmingly logical, concentrating on *why* something is done or must be done, rather than on *how* it is being carried out.

1.8 SUMMARY

This chapter has introduced a number of features of SSADM. It has:

- located SSADM in the systems development lifecycle;
- emphasised its product perspective;
- provided an example Product Breakdown Structure;
- introduced the Structural Model of SSADM;
- described the three-fold modelling perspective;
- introduced logical modelling.

Finally, the context of this book has been re-stated. It focuses on techniques, products and activities and it summarises the development framework (see Chapter 12).

2   Structure Diagrams

2.1   INTRODUCTION

The SSADM Structure Diagram is used in several different SSADM techniques which require similar styles of representation. This chapter describes this notation in general terms to avoid unnecessary repetition in subsequent chapters. The following diagrams utilise this common notation:

Entity Life History	(ELH)
Effect Correspondence Diagram	(ECD)
I/O Structure Diagram	
Dialogue Structure	
Enquiry Access Path	
Update Process Model	(UPM)
Enquiry Process Model	(EPM)

As the notation is used in such a wide number of diverse circumstances, it is particularly important that the context of a structure diagram is clearly understood before it is interpreted.

2.2   DIAGRAM NOTATION

The structure diagram is based upon the concepts of sequence, selection and iteration, the notation used being derived from that used in Jackson Structured Programming. The diagram is drawn as an inverted tree with the top-most node as the root node, representing the object whose structure is being described. This may be an entity whose life history is being modelled, an interface whose structure is being specified or a process whose operation is being defined.

The object being modelled may be considered to comprise several components. These may be grouped as follows:

– as a sequence of components following on one from the other;

– as a selection of components where only one is chosen from the group;

19

– as an iteration where the component may be repeated zero, one or many times.

Each component on the diagram is represented by a box. The diagram is drawn with either round-edged (soft box) or rectangular (hard box) boxes depending upon the context.

The general structure of the notation will be introduced by looking at the life of an employee in some mythical organisation.

2.2.1 Sequence

One of the most common sequences in the life of an entity is the birth-life-death sequence. For an employee the analogous steps might be Hired, Works one day and Dismissal. Each of these components is represented by a box and the sequence is shown in Figure 2.1.

It must be noted that Figure 2.1 states that the only permitted series of events for an employee is Hired, Works one day and Dismissal. The employee in this mythical scenario must work one, and only one, day before being dismissed! Specifically, the node Employee, is a sequence comprising Hired, Works one day and Dismissal. The nodes Hired, Works one day and Dismissal are children of the node Employee, which is their parent node. The nodes Hired, Works one day and Dismissal are siblings of each other. A node on the diagram may have only one parent node but may have many children nodes.

2.2.2 Iteration

The permitted sequence of events for an employee described in Figure 2.1 can only represent a rather unusual working environment. It is likely that an employee may work on zero or many days before being dismissed. Thus there is a requirement to represent a particular component which is repeated zero or many times. This iteration of a component

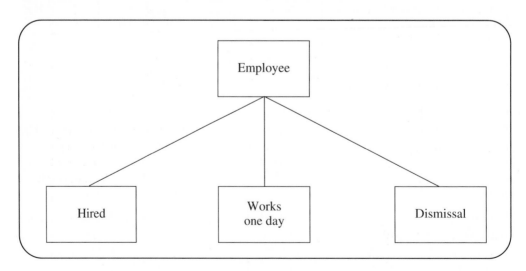

Figure 2.1

is represented by placing an asterisk in the top right-hand corner of the box. In Figure 2.2 the node Work-life is an iteration of node Works one day.

An iteration may have only one child node.

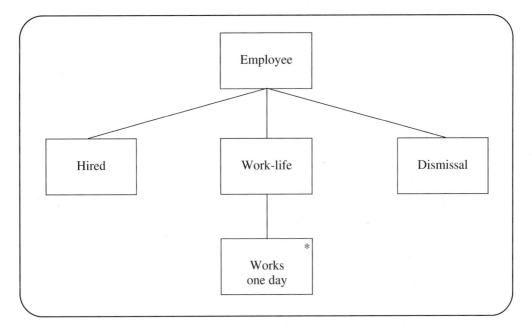

Figure 2.2

Many possible series of events for an employee are described explicitly by Figure 2.2. For instance, the following series of events are all consistent with Figure 2.2:

- Hired, Dismissal;
- Hired, Works one day, Works one day, Dismissal;
- Hired, Works one day, . . . , Works one day, Dismissal;
- Hired, Works one day, Dismissal.

The diagram describes an unlimited number of possible different series of events for an employee but each of these is constrained by the rules specified by the diagram.

2.2.3 Selection

No doubt there are organisations which do not always fire their employees. On some occasions an employee may resign. As a result, an employee life may end in dismissal or resignation. These options can be represented on a structure diagram as the selection node End of career which comprises Dismissal and Resignation. Each element of a selection is represented by placing an 'o' in the top right hand corner of the box. A selection

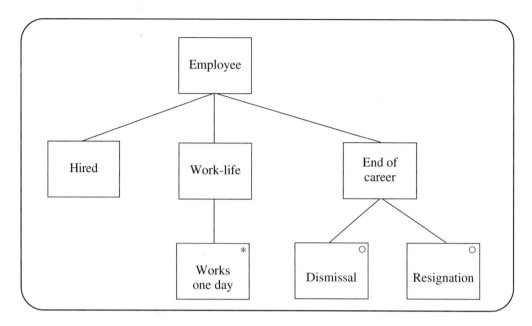

Figure 2.3

must have at least two child nodes. Figure 2.3 shows another structure diagram for employee which includes these additional events.

The structure diagram in Figure 2.3 describes a more elaborate permitted series of events. The following series of events are all consistent with Figure 2.3:

- Hired, Resignation;

- Hired, Works one day, Works one day, Dismissal;

- Hired, Works one day, . . . , Works one day, Resignation;

- Hired, Works one day, Dismissal;

- Hired, Dismissal.

In Figure 2.3 the leaf nodes Hired, Resignation, Works one day and Dismissal are elementary nodes as they are not decomposed any further.

It is important to note that sibling nodes must be of the same type. Figure 2.4 demonstrates an incorrect syntax.

To correct this diagram, the iterated node Works one day must have an iteration node inserted above it. Equally, the selected nodes Dismissal and Resignation must also have a selection node above them.

On occasions, it may be necessary to show that a component may be optionally present. This is achieved by using a selection construct which includes a null component. This null component is represented by a box containing a dash. For example, the employee may be trained before commencing his/her work life. Figure 2.5 illustrates this.

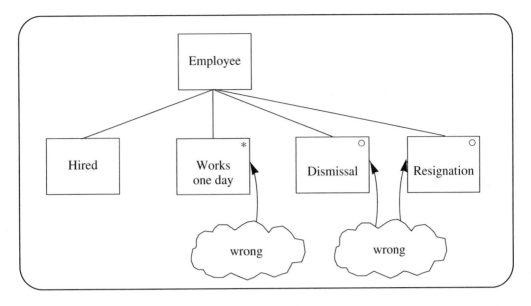

Figure 2.4

The node Training is a selection comprising the node Trained and the null node. The following series of events are all consistent with Figure 2.5:

- Hired, Trained, Resignation;

- Hired, Works one day, Works one day, Dismissal;

- Hired, Works one day, . . . , Works one day, Resignation;

- Hired, Trained, Works one day, Dismissal;

- Hired, Dismissal.

Figure 2.5 hints at one of the difficulties encountered when preparing structure diagrams – that is, the naming of intermediate nodes such as Training or Work-life. Ideally the name chosen should reflect the sub-structure beneath the intermediate node and in Figure 2.5 the naming is relatively straight-forward (though Possible training and Potential work-life might be more meaningful).

When sub-structures are more complex it can be quite difficult to chose meaningful and unique names for the intermediate nodes. This is particularly the case when the intermediate node is introduced to satisfy the demands of the syntax rather than reflecting some naturally existing substructure. Nonetheless, the documentation is more meaningful if all the nodes are named.

2.3   CONSTRUCTING A DIAGRAM

When constructing a diagram it is often necessary to redraw the diagram several times before it accurately describes the structure. Although the final diagram is hierarchical, preparing a diagram is not solely a top-down technique. If a diagram is being constructed

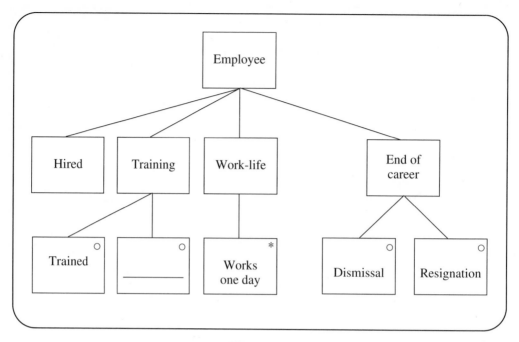

Figure 2.5

from its components, the following steps may be helpful.

- List all the components of the object being described.
- Identify all the groups of alternative components (selection).
- Identify the components or groups of components which are repeated (iteration).
- Identify the groups of components that form sequences.
- Sketch the diagram.
- Check that all the possible series of components described by the diagram are consistent with the rules that govern the structure of the object.
- Check that all legitimate series of component are described by the diagram.
- Redraw the diagram as necessary.

These steps will be applied to the preparation of an Output Data Structure for an enquiry requesting all the orders for a customer. The output for this enquiry comprises:

Customer details;

Order header details;

Ordered bought-in items;

Ordered manufactured items.

The components Ordered bought-in and Ordered manufactured items are identified as

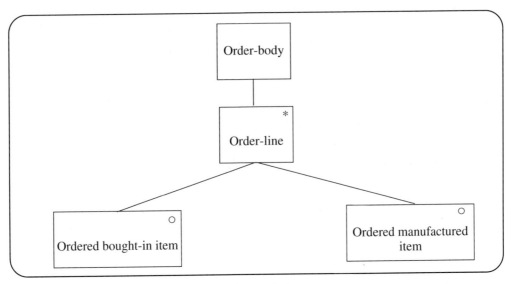

Figure 2.6

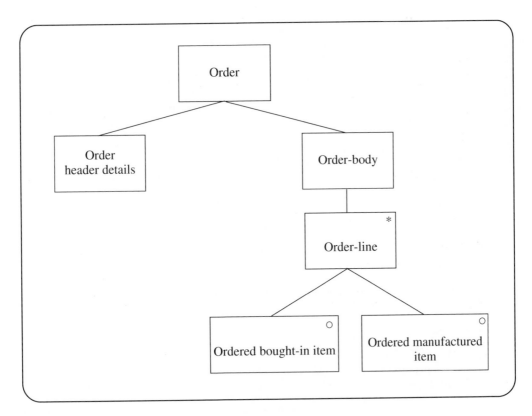

Figure 2.7

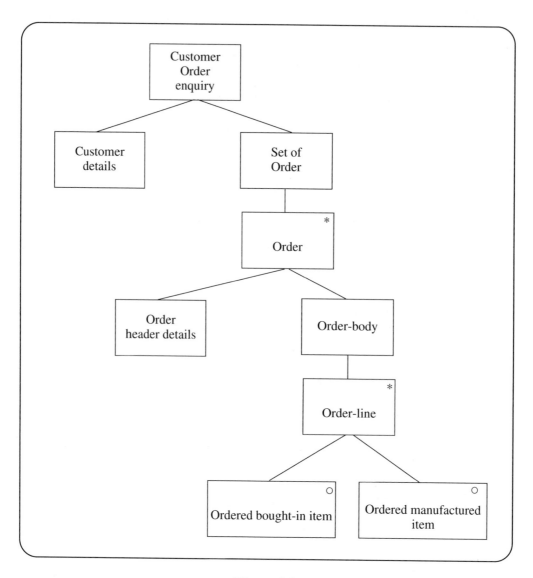

Figure 2.8

alternatives. This group of components is seen to repeat many times giving the sub-structure shown in Figure 2.6.

The components Order header details and Order body are now identified as a sequence, which again iterates giving the diagram in Figure 2.7.

Finally it is identified that the component Order is repeated and that this repetition is in sequence with Customer details. This gives the final structure shown below in Figure 2.8.

This diagram could have been developed using a top-down decomposition technique rather than the bottom-up technique used. Neither approach is always superior, though

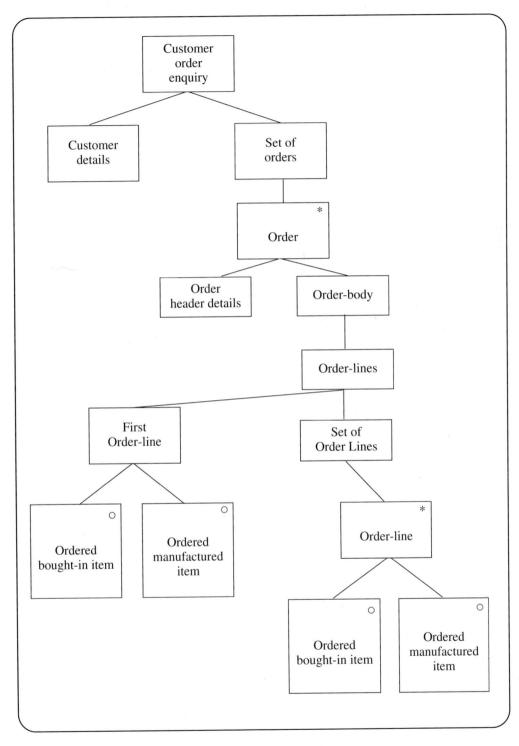

Figure 2.9

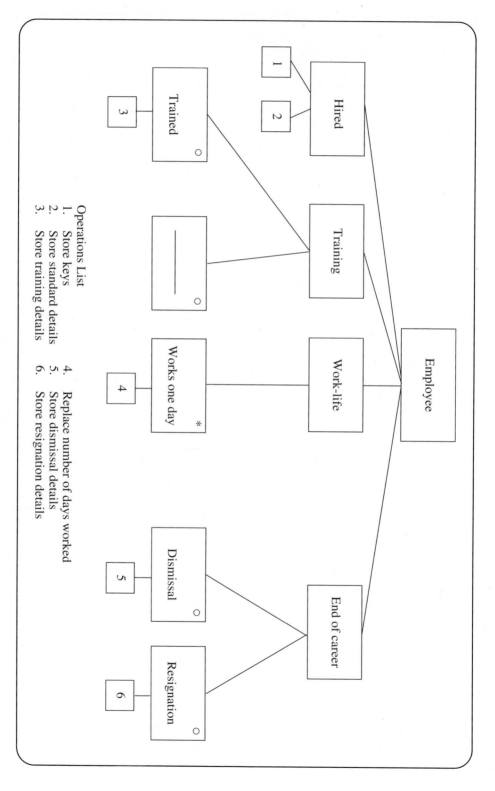

Figure 2.10

SSADM users may find the top-down technique easier for the relatively straightforward parts of a structure and the bottom-up technique useful for the more complex. Structure diagrams may also be derived from other structure diagrams in SSADM. For instance, the process model may be based on structure diagrams which describe the input and output data structures for that process.

Figure 2.8 is consistent with orders having no order-lines. This may, of course, accurately describe the business rules but, if it does not, then the diagram should be redrawn. Figure 2.9 shows the redrawn diagram which reflects the rule that an order must have at least one order-line.

2.4 FURTHER NOTATION

A: Operations

When the structure diagram is used to describe Entity Life Histories or Process Models, the nodes may be further detailed with their constituent operations. This is appropriate as the nodes on these diagrams represent the effects of events or the processes which cause the effects of events. The types of allowable operation are specified in subsequent chapters but in general the operations are concerned with database access. Figure 2.10 shows the Employee structure diagram with appropriate operations added. The operations are numbered, and numbered boxes are drawn below the relevant elementary component of the structure diagram. Note that the operations below a node are applied in the sequence shown on the diagram, and should be considered as a more detailed description of the operation of the node rather than sub-nodes in themselves.

B: Parallelism

This construct is only used in Entity Life Histories. On occasions, permitted orderings of the components of a structure may not be adequately described by the constructs of sequence, iteration and selection. For instance, an employee may change address or marital status independently of the other events shown in Figure 2.5.

From the perspective of a personnel system, change of address and change of marital status may happen repeatedly and in parallel with the group of events, Trained and Works one day. A node that comprises a series of parallel nodes is separated from its children by a parallel bar. An illustration is shown in Figure 2.11. Note that Change address and Change marital status are iterations as they may occur zero or many times.

The following series of events are all consistent with Figure 2.11.

- Hired, Trained, Change address, Resignation;

- Hired, Change address, Trained, Resignation;

- Hired, Works one day, Change address, Works one day, Dismissal;

- Hired, Works one day, ..., Works one day, Resignation;

- Hired, Change address, Trained, Change marital status, Works one day, Change

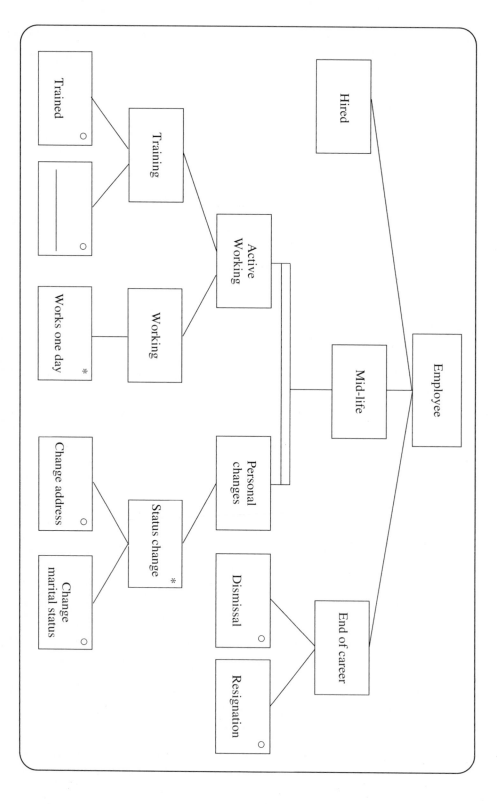

Figure 2.11 Parallelism

address, Change marital status, Dismissal;

- Hired, Dismissal.

2.5   USES IN SSADM

In SSADM Version 3 the structure diagram was only used for the Entity Life History. In Version 4 this notation is used much more widely and is applied during the following activities:

Function Definition;

Entity-Event Modelling;

Logical Data Modelling;

Dialogue Design;

Logical Database Process Design;

Physical Design.

The development of one type of structure diagram from another is shown by the illustration in Figure 2.12.

In Physical Design the notation may also be used to specify the Program Data Interface if required.

2.6   SUMMARY

The structure diagram is based upon the constructs of sequence, iteration and selection. When the notation is used for Entity Life Histories or Process Models, operations may also be added to the diagram. If the notation is being applied to Entity Life Histories, the parallel construct may also be used. The construction of a diagram is governed by the following rules:

- the root must comprise either a sequence, a selection or an iteration;
- all boxes below the root must have exactly one parent;
- all children of any one parent must be of the same type;
- each box containing the iterate symbol must be the only child of its parent;
- a selection must comprise at least two choices, one of which may be the null box;
- every parallel component must be part of a sequence
- every parallel bar must have at least two boxes immediately below it and none may contain iterate or selection symbols.

A structure diagram may be developed in either a top-down or bottom-up fashion or it may be derived from other structure diagrams.

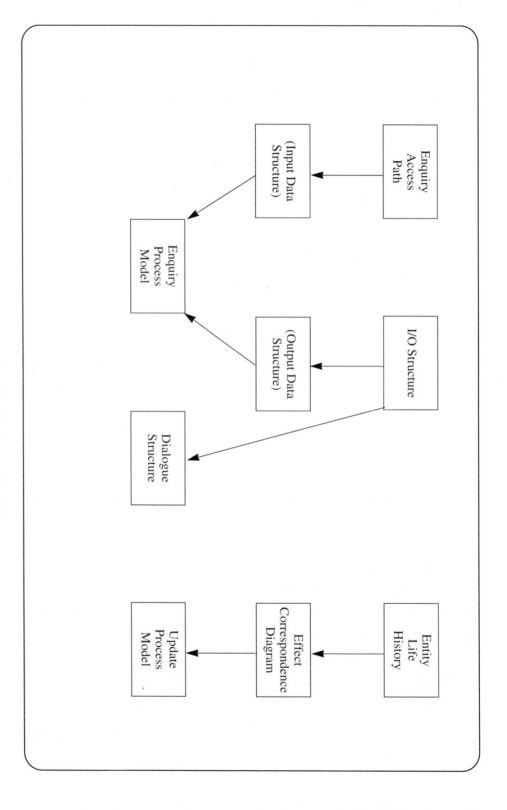

Figure 2.12 Derivation of Structure Diagrams

3   Data Flow Model

3.1   INTRODUCTION

Data Flow Diagrams (DFD) show the passage of data through the system. They focus on the processes that transform incoming data flows (inputs) into outgoing data flows (outputs). The processes that perform this transformation normally create and use data (data stores). External entities send and receive data flows from the system.

Most users perceive systems from a process perspective and this is reflected in their descriptions of the systems during fact finding.

"We receive these day sheets and amalgamate them into a regional summary."

"The timber is available for dispatch as soon as it has cleared customs."

Consequently, a graphical process model such as the Data Flow Diagram is normally very accessible to users and operators of the system. Its focus on transformations closely reflects how they themselves perceive the system.

SSADM uses Data Flow Diagrams (DFD) in three stages of the development process:

– Current Physical DFDs. These record the results of conventional fact finding.

– Current Logical DFDs. The logical information processing of the current system.

– Required Logical DFDs. The logical information processing requirements of the proposed system.

3.2   MODELLING NOTATION

The Data Flow Diagram depicts the passage of data through a system by using five basic constructs.

3.2.1 Data Flows

A data flow is a route which enables data to travel from one point to another in the diagram. The flow is shown as an arrowed line with the arrowhead showing the direction of flow. The flow is given a simple meaningful descriptive name.

Order Details

$\longrightarrow$

Flows may move from an external entity to a process, from a process to another process, into and out of a store from a process and from a process to an external entity. Flows are not permitted to move directly from an external entity to a store or from a store directly to an external entity. In these instances further processes must be defined to store incoming data flows or create outgoing ones.

It is generally unacceptable to have a flow moving directly from one external entity to another. However, if it is felt useful to show these flows, and they do not clutter the diagram, they can be shown as dotted lines.

In logical DFDs, no two data flows should have the same name. If two flows, separated by a process, appear to be identical then it is likely that the process that separates them is doing nothing and should be discarded. It is also useful to adopt this unique naming convention for physical DFDs as it leads to a tighter definition of the current system. The data flows moving in and out of stores do not necessarily require names because the store name may be sufficient to describe them. However, in some instances it may be useful to use a name where the flow is especially significant or is not easy to discern from an examination of the diagram.

It may be possible to give a combined name for circumstances where many flows move between the same source and destination.

For example, the following flows:

   Weekly aged creditors report;

   Weekly aged debtors report;

   Monthly trial balance;

   Monthly account listing;

might be best described as:

   Financial reports.

3.2.2 Processes

Processes are transformations, changing incoming data flows into outgoing data flows. A rectangular box denotes a process. All processes are numbered to permit easy identification. Although the numbering generally follows a left to right convention, there should be no rigorous attempt to show priority or sequence, particularly in logical diagrams.

The name of the process should describe what happens to the data as it passes through it. A suggested notation is an active verb (extract, compute, verify) followed by an object or object clause. For example:

```
┌─────────────┐
│   Verify    │
│   Credit    │
│   Status    │
└─────────────┘
```

Other active, unambiguous verbs include create, produce, retrieve, store, determine and calculate. The word process should be excluded from any name. Not only is it an ambiguous term, but also the notation of the box already conveys this information.

In physical Data Flow Diagrams the location of the process may be included in the stripe at the top of the process symbol. No physical activities or arrangements should intrude into the process description of logical DFDs. 'Sort records into alphabetical sequence' is not a logical function. Similarly, 'Accounts Clerk extracts Supplier Code' is also not allowed because it represents a physical commitment to how the operation will be carried out. In general, physical references are one of the following:

- Organisational: Director signs Requisition Note;

- Procedural: Sort and Batch Invoices;

- Redundant: Copy Invoice to Logistics.

3.2.3 Data Stores

A store is a repository of data; it may be a card index, a database file or a folder in a filing cabinet. The store may contain permanent data (such as a sales ledger) or temporary accumulations (pending documents, daily movements).

Each store is given a meaningful descriptive name. Data stores may be included more than once to simplify the presentation of the Data Flow Diagram. Such multiple occurrences are shown by an additional vertical line within the store symbol. Each store is also given a reference number prefixed by a letter D to represent a computer data store or M for a clerical data store.

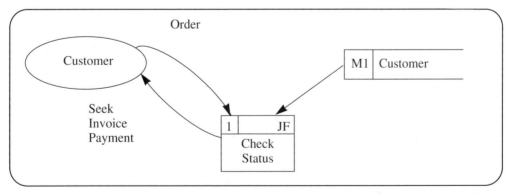

Figure 3.1   Process does not alter contents of store

In the example given in Figure 3.1, the arrow from the store is single-headed and points towards the process. This is to signify that the process does not alter the contents of the store, it only uses the data available. However, if the contents of the store are altered by the process, as well as being read, then the diagram uses a double-headed arrow.

In this way a single-headed arrow shows READ (looking at the data only) or WRITE (changing data only) operations. The creation of the store 'Hold' forms is a WRITE-only function of the Check Status process (see Figure 3.2). In circumstances where the data is both examined and changed (READ and WRITE) then two separate flows are shown. However, to avoid an excessive number of data flows on a high level diagram, double-headed arrows may be used.

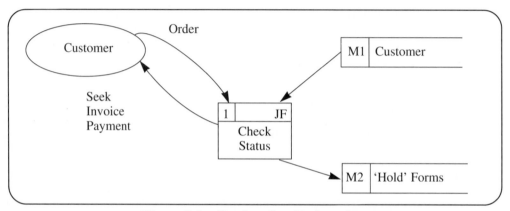

Figure 3.2   Read and write functions

3.2.4 External Entities (Source or Sink)

A source or sink is a person or part of an organisation which enters or receives data from the system but is considered to be outside the scope of the project. The source or sink may be duplicated in a completed Data Flow Diagram to simplify presentation. This duplication is again shown by the addition of a line within the symbol. External entities

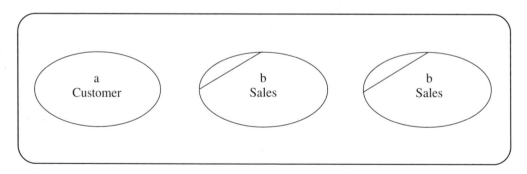

Figure 3.3   External entity and duplicated External entities

may be further referenced by the use of an alpha character, and this is particularly recommended if a lower level diagram decomposes the entity.

Because they are considered to be outside the scope of the project, external entities force continual review of the boundaries of the system. As analysis progresses it might be necessary for this boundary to be re-assessed and so certain sources or sinks are brought into the agreed scope of the project. Figure 3.3 demonstrates the notation.

	External Entity	Process	Data Store	Resource Store
External Entity	External data flow and resource flow only	YES (data flow & resource flow)	NO	NO
Process	YES (data flow & resource flow)	YES (data flow & resource flow)	YES	Resource flow only
Data Store	NO	YES	NO	NO
Resource Store	NO	Resource flow only	NO	NO

Data flows NEVER join up to each other, nor do resource flows

Figure 3.4   Permitted connections of DFD components

3.2.5 Physical Resources

A broad arrow is used to show the flow of physical resources and a closed rectangle to represent resource stores. These symbols are not widely used on Data Flow Diagrams because of the clutter produced by their physical size. However, they can be useful for:

- showing significant resource flows and states. This representation is often more meaningful to users than logical data flows which may appear a little abstract.

- getting started in a project. Users may describe the system in terms of physical flows and stores. The construction of a flow diagram using these symbols permits the identification of the information flows which are needed to support and record the physical movements of goods and materials.

– finally, and importantly, the physical resources may actually be what the system is about! It is certainly equally important to send both the goods and the invoice. The practical aspects of the system may be lost to an analyst who concentrates too much on the neatness of the data flow.

The permitted connections between DFD components are summarised in Figure 3.4.

3.3   MODELLING HIERARCHY

The Data Flow Diagram has a simple and consistent way of representing the successive modelling levels required in a hierarchy. Each process is exploded into a lower level Data Flow Diagram until the process can be represented in an Elementary Process Description. In this way, it is possible to present a series of Data Flow Diagrams representing increasing levels of detail, and appropriate for different types of task and staff. The convention of denoting the highest level diagram as level 1 has been adopted with subsequent levels designated 2, 3, etc. Indeed, the concept can also be extended backwards where the complete level 0 Data Flow Diagram is a one process diagram which summarises the inputs and outputs of the system under consideration. This is called the context diagram.

The numbering system of the Level 1 model needs to be extended to the lower level diagrams so that they can easily be referenced back to their 'parent' processes. This is achieved by using a decimal numbering system. Thus process 3 may be decomposed into 3.1, 3.2, 3.3, etc and, if a further level is required, 3.1 into 3.1.1, 3.1.2, 3.1.3, etc.

It is important to check that all flows and stores accessed by the process in the higher level box are actually used in the lower level decomposition. Data Flow Diagrams should be balanced so that data flows in and out of a process must appear on the Data Flow Diagram that is the decomposition of that process. If a flow does not appear then the reason for its existence at the higher level must be closely examined. Similarly, flows should not be produced that do not exist in the parent process. If such a flow seems to be necessary then all higher level diagrams should be altered so that the whole model remains balanced.

It is also possible to break down external entities and data stores in a lower level diagram. However, in such instances the notation must be closely adhered to so that cross references to the higher level diagram can be clearly identified.

Three general guidelines can be given for controlling decomposition of Data Flow Diagrams.

Decomposition is complete when:

– All processes on the lowest level DFD can be described in an A4 Elementary Process Description.

AND

– There is no process on the lowest level DFD with a read/write flow to a store (because these are usually undertaken by two separate processes).

AND

– An exploratory interface line (showing the human/computer boundary) does not intersect any processes on the lowest level DFD.

Most systems can be modelled within two or three levels. Processes decomposed beyond this often become trivial and do not access stores in any way.

The lowest level of decomposition of any process is shown by an asterisk on the bottom right-hand corner of the box. Obviously such processes can occur on any of the levels of the diagram. It is quite feasible for processes on the level 1 DFD not to require further decomposition.

3.4 CONSTRUCTING THE DATA FLOW MODEL

This section examines the construction of a simple Data Flow Diagram and illustrates its progression from a representation of the current physical system through current logical to required logical.

3.4.1 The Current Physical System

Figure 3.5 records an interview between the analyst and a sales clerk. The essential operations are summarised below:

Process:	Log Sales Enquiry
Performed by:	Sales Clerk
Inputs:	Sales Enquiry
Outputs:	Blue Enquiry Form

Process:	Split and File Enquiry
Performed by:	Sales Clerk
Inputs:	Blue Enquiry Form
Outputs:	Copy-1 of the Blue Enquiry Form
	Copy-2 of the Blue Enquiry Form

Process:	Allocate Enquiry
Performed by:	Sales Manager
Inputs:	Copy-2 of the Blue Enquiry Form

Outputs: Allocated Enquiry

Process: Pursue Enquiry

Performed by: Product Manager

Inputs: Allocated Enquiry

Outputs: Enquiry Result

Process: Complete Enquiry Form

Performed by: Product Manager

Inputs: Enquiry Result

Outputs: Completed Enquiry Form

Process: Enter and File Sales Result

Performed by: Sales Manager

Inputs: Completed Enquiry Form

Outputs: PC-file

                         Copy-2 Enquiry Form

Process: Compile Weekly Report

Performed by: Secretary

Inputs: Copy-2 Enquiry Form

Outputs: Weekly Report

INTERVIEW RECORD

SYSTEM NAME EXAMPLE	DATE 01.01.92	TIME 14.00	DURATION 30 mins	AGENDA AG 11
PARTICIPANTS Sales Manager's Clerk Systems Analyst				AUTHOR SA

| INTERVIEW OBJECTIVES
To understand how sales enquiries are administered | |

RECORD	X-REF
All telephone sales enquiries are logged by the sales clerk on a blue enquiry form. A copy of this two part form is filed in the sales office. The other copy is sent to the Sales Manager who allocates the enquiry to the appropriate Product Manager. The Product Manager pursues the enquiry until either • A sale(s) is made • No sale is made and the lead is deemed to be "dead" • The lead is deemed to be dormant and will be chased again at a later date. In the first two instances the copy of the blue enquiry form is returned to the Sales Manager with the ⁄ contd.	CD1

Figure 3.5   Interview record

INTERVIEW RECORD

SYSTEM NAME EXAMPLE	DATE 01.01.92	TIME 14.00	DURATION 30 mins	AGENDA AG 11
PARTICIPANTS Sales Manager's Clerk Systems Analyst				AUTHOR SA

INTERVIEW OBJECTIVES See page 1	
RECORD	X-REF

result of the lead completed. He enters the form details on to a PC and files the form.

At the end of each week the Sales Manager's secretary goes through the copies in his file and reports on

- Number of enquiries received
- Number resulting in a sale
- Number not resulting in a sale

The report is sent to all regional sales managers.

Figure 3.5   (Cont.)

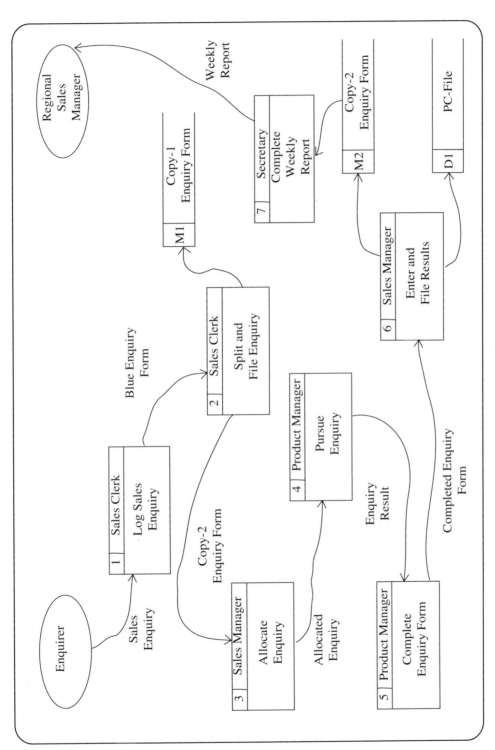

Figure 3.6 First-shot DFD for sales enquiry system

Figure 3.6 is a possible representation of this narrative in a Data Flow Diagram.

This first attempt raises a number of general issues:

- Examine the usefulness of processes that do not create or access stores. For example, processes 1 and 2 could be combined without losing information.

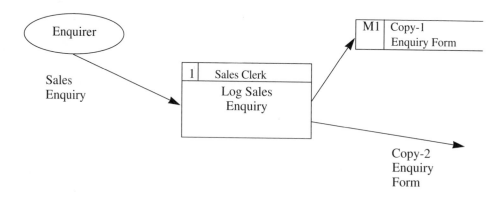

The filing of the enquiry is explicitly shown in the data flow to the store. The multi-copy nature of the form is implicit in the naming of the data store.

- Check that processes not connected to stores do not access data in some way. For example, how is the allocation achieved in process 3? On further investigation it is revealed that the Sales Manager examines a sales product list and this is added to the DFD.

- Check to see whether a process creates and uses a temporary file before it is resolved. Process 4 is a candidate for this. The following is more accurate.

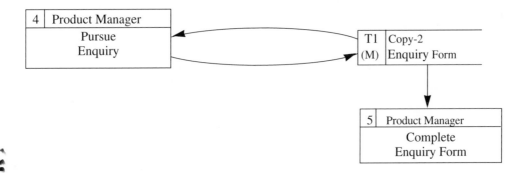

With process 5 accessing the data of the store.

- Examine sources and sinks. It may be more acceptable to make the PC an external entity rather than a store.

  This gives a re-drafted DFD which is more accurate and gives a better balanced view (Figure 3.7). This is a reflection of the current physical system.

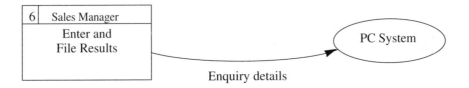

3.4.2 The Current Logical System

Re-casting the current physical Data Flow Diagram into its logical equivalent generally requires four tasks:

- rationalisation of data stores;
- rationalisation of bottom level processes;
- grouping of new bottom level processes into higher level processes;
- walkthrough checks for consistency and completeness.

The first two of these may be illustrated by the simple sales enquiry system.

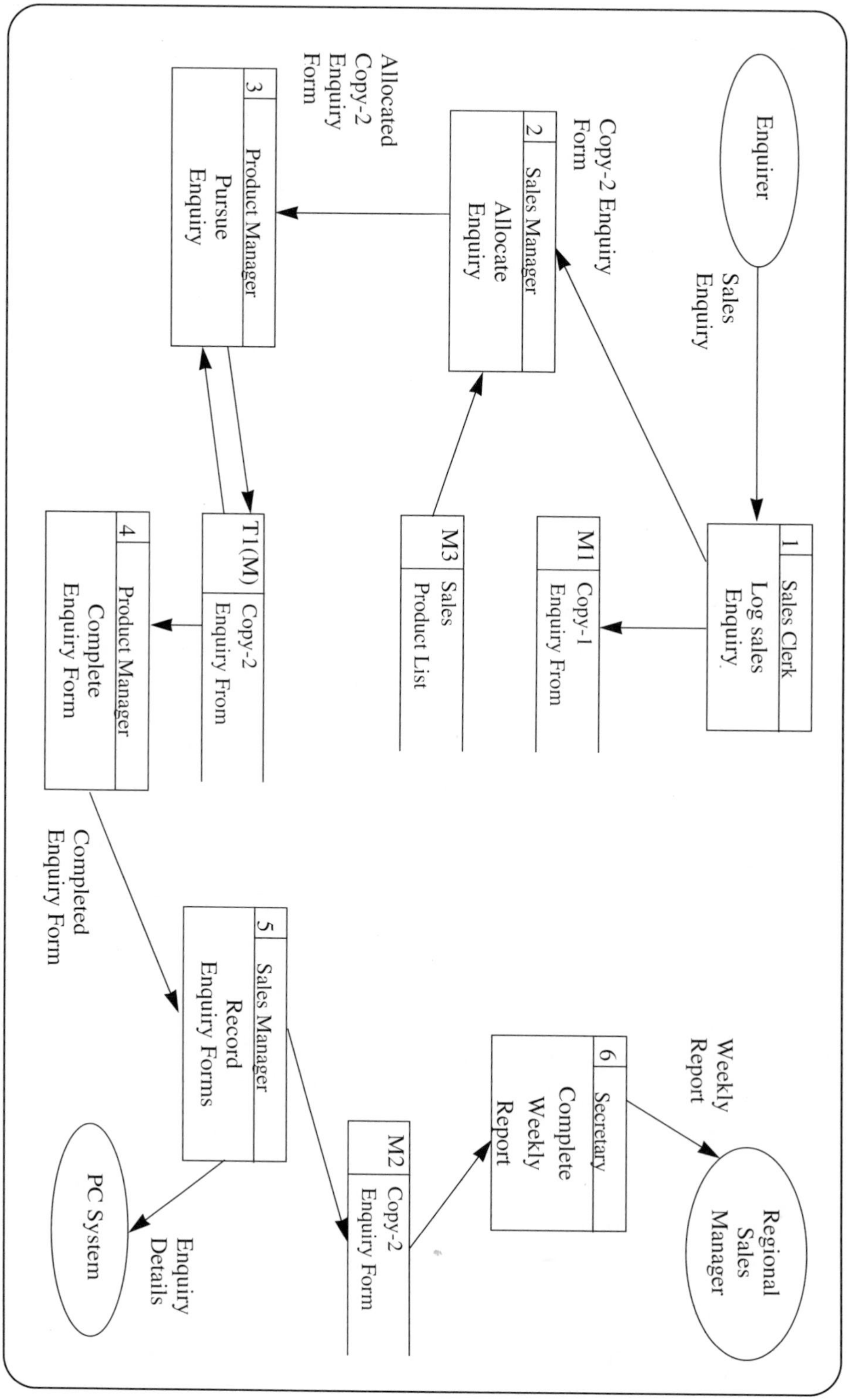

Figure 3.7   Re-drafted current physical DFD for sales enquiry system

3.4.3　Rationalisation of Data Stores

This is generally achieved by merging data stores which have identical keys or identifiers.

In this example the data stores are as follows:

M1 Copy-1 Enquiry Form	identifier: enquiry-no
	logical store: Enquiry
M3 Sales Product List	identifier: product-no
	logical store: Product
M2 Copy-2 Enquiry Form	identifier: enquiry-no
	logical store: Enquiry
T1(M) Copy-2 Enquiry Form	identifier: enquiry-no
	logical store: Enquiry

Thus the current logical DFD will only show two logical stores: Product and Enquiry.

3.4.4　Rationalisation of Bottom Level Processes

General rules about the rationalisation and grouping of processes in Data Flow Diagrams are difficult to give. A number of possibilities exist and will be applicable in different circumstances. For example:

- Remove 'reorganising' processes such as Sort Locations or Index Applications.
- Remove 'retrieve only' processes and record in the Requirements Catalogue.
- Remove processes which will be dealt with outside the system.
- Combine processes which are essentially a series of activities.

Each application must be dealt with on its merits. These may be assessed in walkthroughs which check for consistency, completeness and clarity.

If the current physical DFD has been drawn at an appropriate level then it is likely that most processes will progress unchanged to the logical stage. However, the current terminology needs examining to see if there are any undue references to the physical nature of the activity. Slight changes may result. For this example:

Old process name	**New process name**
1 Log Sales Enquiry	1 Create Sales Enquiry
2 Allocate Enquiry	2 Allocate Enquiry
3 Pursue Enquiry	3 Pursue Enquiry
4 Complete Enquiry Form	—
5 Record Enquiry Results	4 Record Enquiry Results

6 Compile Weekly Report   5 Compile Weekly Report

Process 4 does not access a permanent store and is solely an element of procedure. Consequently, it is dropped from the logical diagram. However, enquiry details still have to be available and this is now logically shown as a data store access by the new process 4 (Record Enquiry Results). This is equivalent to the access required by process 4 of the physical DFD.

The results of logicalisation are shown in the current logical DFD of Figure 3.8. Note that all physical locations have been removed from processes and data flow names no longer refer to physical media.

3.4.5   Required Logical System

The required logical Data Flow Diagram will reflect new user requirements. New processes will be defined to reflect new needs, whilst activities which are no longer required or are dealt with elsewhere in the system will be discarded.

For the sales enquiry system, the following requirements have been established:

- the ability to cross-reference enquiries to companies and to produce a report showing enquiry analysis by company;

- to stop the duplication of effort caused by re-entering data into the PC system.

The current logical DFD also has a loose end. It is still unclear how product managers are allocated to products.

An initial required logical DFD is produced (see Figure 3.9) which has:

- added a new Company store to hold stored data required in reports;

- introduced a new process (process 9) to produce this report;

- added two new processes to deal with product creation, amendment and assignment. These have resulted from a better understanding of this aspect of the system gleaned from further user interviews;

- deleted outputs no longer required.

This diagram is now becoming cluttered and re-structuring is required.

Furthermore, two further issues need addressing:

- Create Sales Enquiry appears to contain the creation of Company details when an enquiry is received from a company not currently on file. Thus this process box probably contains the procedures required to:

  create a sales enquiry;

  create a company record;

  amend company details;

  allocate a company to an enquiry.

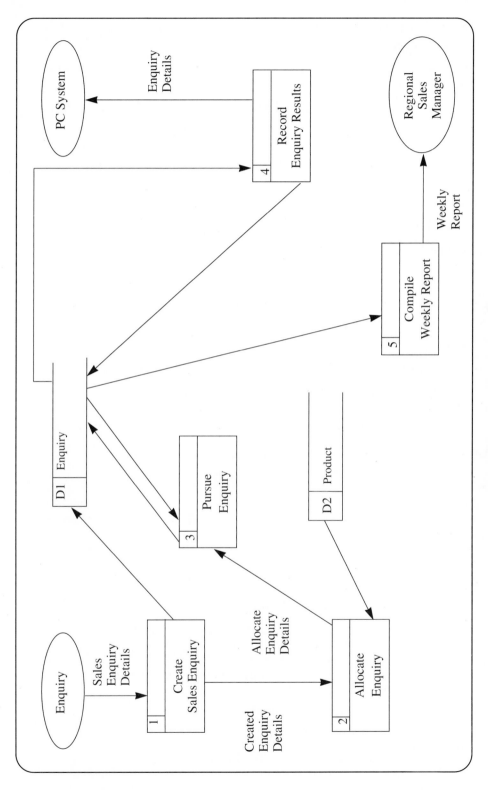

Figure 3.8   Current logical DFD for sales enquiry system

These could be entered on the DFD but would raise the number of processes to 13 in an already cluttered diagram.Thus a hierarchical set of DFDs might be preferred.

    – The trigger for Record Enquiry Results is unclear. The following clarification has been received from the user in a subsequent interview.

> The receipt of an order from a customer means that a sale has been made.
>
> If an order is not received within six months of the enquiry then the enquiry is dead unless authorisation is received from the Sales Manager to keep the enquiry 'live'.

The required logical DFD is now re-drawn to reflect these points (Figures 3.10 and 3.11).

3.5   FURTHER NOTATION

3.5.1 Data Stores

On lower level DFDs it is possible to show data stores that are only accessed by processes decomposed from the higher level . It is also possible to decompose the data stores of the higher level diagram into constituent parts. For example:

D1	Order

is a data store on the Level 1 DFD which may be decomposed to

D1a	Regular Order

D1b	Standing Order

on Level 2 diagrams.

The following data store only occurs on the Level 2 DFD which details the second process of the Level 1 DFD.

D2/1	On-hold Order

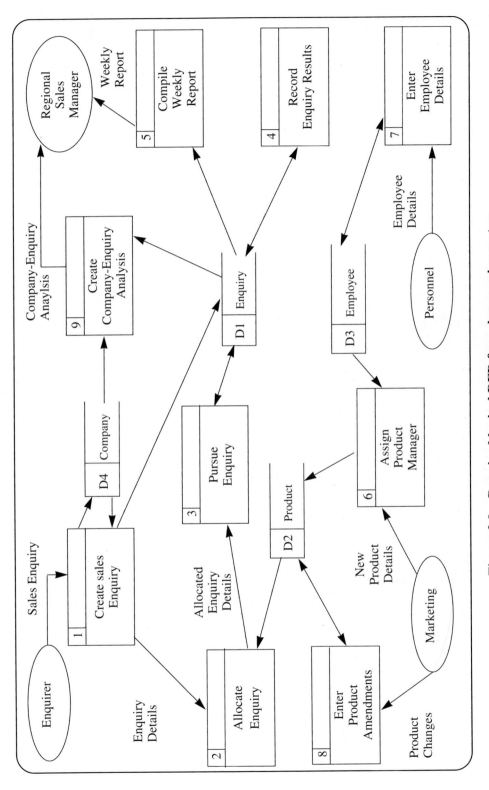

Figure 3.9   Required logical DFD for sales enquiry system

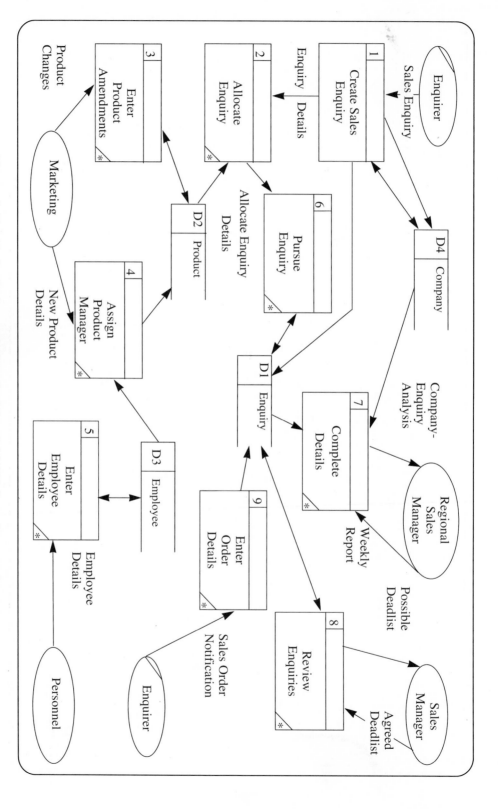

Figure 3.10   Redrafted required logical Level 1 DFD for sales enquiry system

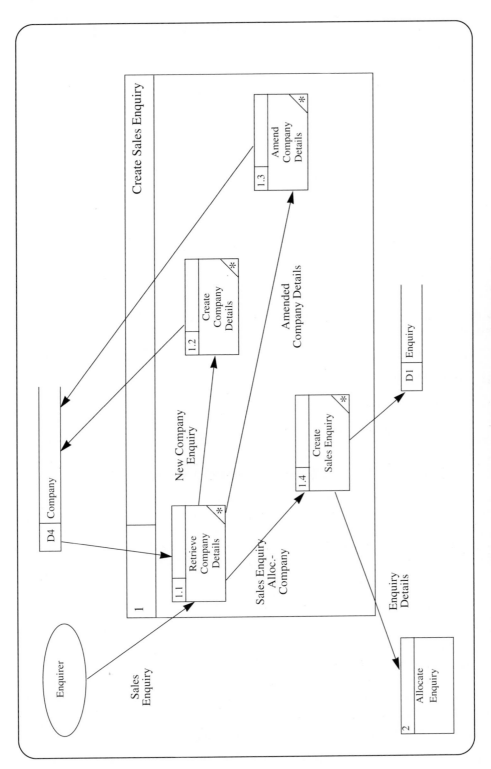

Figure 3.11   Level 2 logical DFD for sales enquiry system

It is also possible to distinguish transient data stores in both manual and computerised systems. These stores represent the temporary holding of data before it is used by a a process and subsequently deleted.

Such files may be found in a current physical system:

T3(M)	Photocopied Issue List

or on a proposed physical system:

T4	Sorted Orders

An example of a temporary store was given in the sales enquiry system (Figure 3.7).

3.5.2 External Entities

External Entities may be uniquely identified by a lower case alphabetic character and this can be decomposed on a lower level diagram:

a   Management

a1   Sales Manager

a2   Product Manager

a3   Marketing Manager

3.6   CONSTRUCTING THE MODEL – TOP DOWN APPROACH

Section 3.4 described the incremental development of a simple Data Flow Diagram. This section gives a more general approach and illustrates further issues arising from a more complex example. Although the example refers to the construction of a current physical Data Flow Diagram, its principles can also be applied to the development of a logical model in circumstances where no current physical system exists.

Stage 1: Establish the major inputs and outputs of the system.

> Establish the source of each input and the receiver of each output and make these external entities.

The major inputs and outputs of the system are likely to be the most significant data flows to the user and hence should be discovered early in the investigation. These are likely to be the main outputs of the system together with the incoming data flows that fuel these outputs. These flows may be shown on a one process context diagram.

This diagram for a delivery scheduling system is shown in Figure 3.12.

Stage 2: Establish a process that handles each input on its arrival into the system.

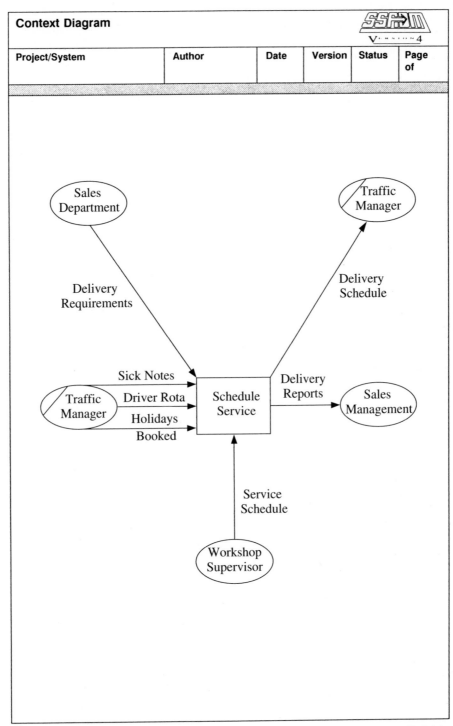

Context Diagram					SSADM Version 4
Project/System	Author	Date	Version	Status	Page of

Figure 3.12   Context diagram for delivery scheduling

SSADMv4 3

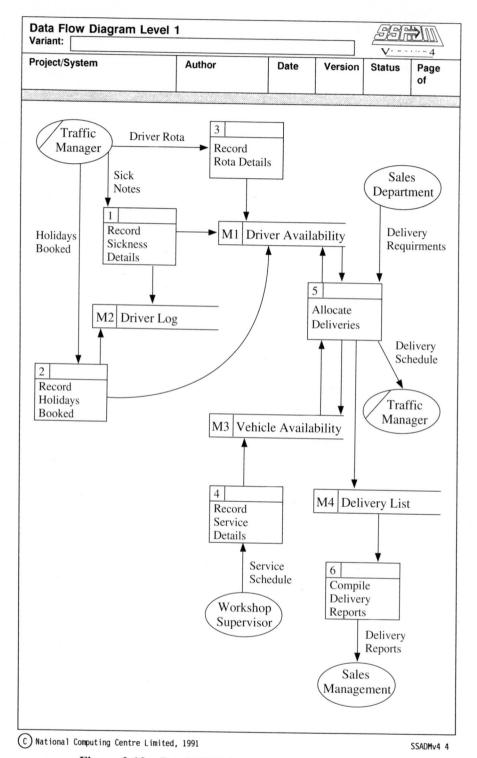

SSADMv4 4

Figure 3.13   Draft DFD for the Delivery Scheduling

Establish the data stores that this process needs to create or use if it is to be successful.

For example, Figure 3.13 shows the arrival of Sick Notes into the system. Analysis must reveal how the system deals with these. An interview with one of the Operations Clerks reveals that Sick Notes are stored in the driver log and the driver availability chart is altered to show that the driver is unavailable because of sickness. This is shown in process 1 with stores M1 and M2 updated by sickness information.

Stage 3: Establish a process that generates each output flow from the system.

Establish a store required by this output for this flow to take place.

Figure 3.13 again illustrates the application of these principles. For example, the delivery reports must be produced from some stored information. On investigation it is revealed that the Delivery List is used and this is reflected in the diagram with process 6 reading the Delivery List store to produce the required reports.

In general processes dealing with inputs will *create* data stores, while processes producing outputs will *use* or *read* data stores. These latter data stores must be created and so further processes must be defined to allow this creation.

Stage 4: Examine the process(es) and associated flows that link the input and output processes.

Establish the rules for these processes and expand the Data Flow Diagram to reflect these rules.

The previous Data Flow Diagram can be checked for completeness with the user. This will allow the main inputs and outputs to be verified. Discussion must then focus on the major *transformation* processes. In this example it is process 5 – allocate deliveries. The current definition looks simplistic and incomplete.

Figure 3.14 reflects the result of further interviews and investigation.

The allocation process has been sub-divided to reflect new information. It transpires that only certain vehicles can be used to deliver to some locations and so vehicle allocation has been split from driver allocation to reflect this. The use of temporary drivers and sub-contractors has also been uncovered and these have now been included in the diagram.

A simple walkthrough of Figure 3.14 also demonstrates that it is incomplete. In fact, it is an abandoned Data Flow Diagram! For example, there are no creation processes for the stores vehicle, location and product. Whilst the processes 'request sub-contractors' and 'request temporary drivers' also appear to be incomplete, it might be expected that these processes should create or update stores of some kind.

Indeed, the main reason for the abandonment of this dataflow diagram was the growing number of processes and stores. It is beginning to lose coherence and clarity.

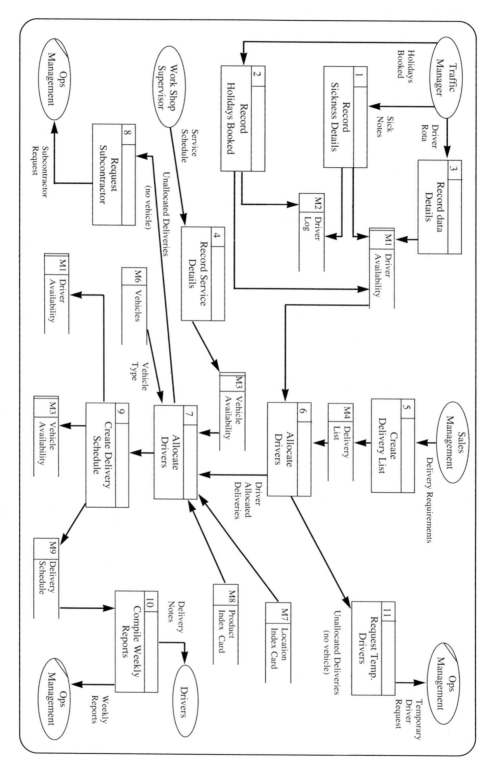

Figure 3.14   An abandoned DFD of delivery scheduling

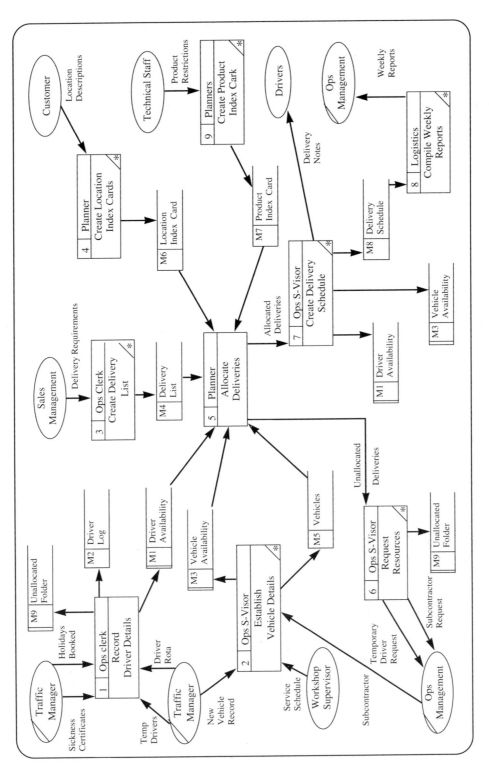

Figure 3.15(a) Level 1 diagram for delivery scheduling

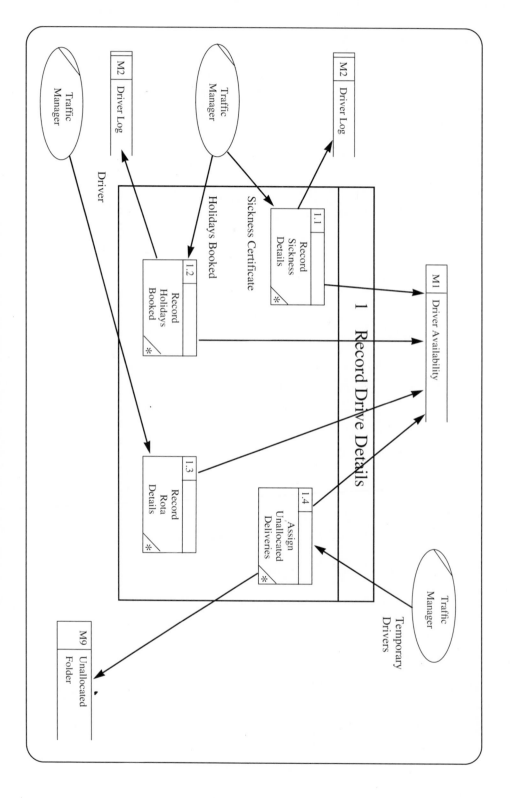

Figure 3.15(b)   Level 2 DFD for process 1 of Delivery Scheduling System

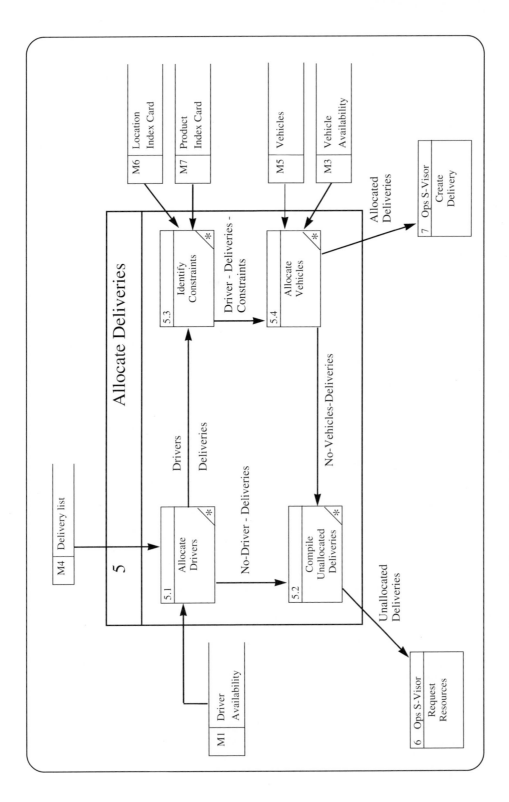

Figure 3.15(c)   Level 2 DFD for process 5 of delivery Scheduling System.

Stage 5: Establish a hierarchical set of Data Flow Diagrams which reduces the number of processes in the top level diagram to a manageable number (say 7 to 10).

Figure 3.15 shows the application of this principle.

Process 1: Record Driver Details and process 5: Allocate Deliveries are now exploded in lower level Data Flow Diagrams. The physical location of these activities has now been added to the diagrams.

This section has shown an incremental approach to Data Flow Diagram development with the model being used as an active analysis tool, prompting the discovery of further inputs and outputs. However, thorough document analysis may have produced a complete diagram of the major data flows between the sections or personnel of the company (see Figure 3.16). This is a Document Flow Diagram and it may have a suitable boundary superimposed on it to produce an appropriate one process model (see Figure 3.17). This document flow summary may be useful at any stage of analysis because it represents an accessible description of the current information system.

3.7   DOCUMENTATION

Data Flow Diagrams provide the most important element of the Data Flow Model (DFM) of the system. This DFM consists of:

- A context diagram providing an overall view of the system.

- A hierarchical set of Data Flow Diagrams.

- Elementary Process Descriptions. One of the guidelines for the decomposition of DFDs was that it was complete when a bottom level process could be defined on a side of A4. An appropriate modelling method for this is discussed in section 3.9.

- External Entity Descriptions. These record information about the responsibilities and functions of the external entity. Figure 3.18 shows an example.

- Input/Output structure Descriptions. These describe the data item content of each data flow that crosses the system boundary. An example is given in Figure 3.19.

3.8   RELATIONSHIP WITH OTHER MODELS

3.8.1   Logical Data Model

The Data Flow Diagrams are constructed in parallel with the Logical Data Model (see Chapter 4) and the two models have to be consistent with each other. This is primarily achieved by linking the stored data of the logical DFDs (the stores) with the stored data of the LDM (the entities with their supporting entity descriptions). This is achieved in a Data Store/Entity cross-reference which is examined in detail at the end of Chapter 4. This parallel construction of the two models means that entities may provide a guide to the logical stores defined in the logical DFDs. So, for example, in the simple DFD constructed in section 3.3, the logical stores Product and Enquiry may have already been suggested by entities of the LDM.

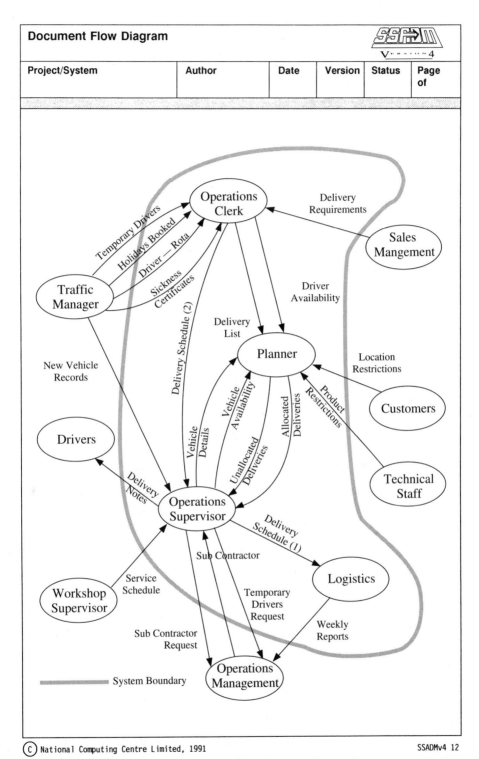

Figure 3.16 Document Flow Diagram for Delivery Scheduling

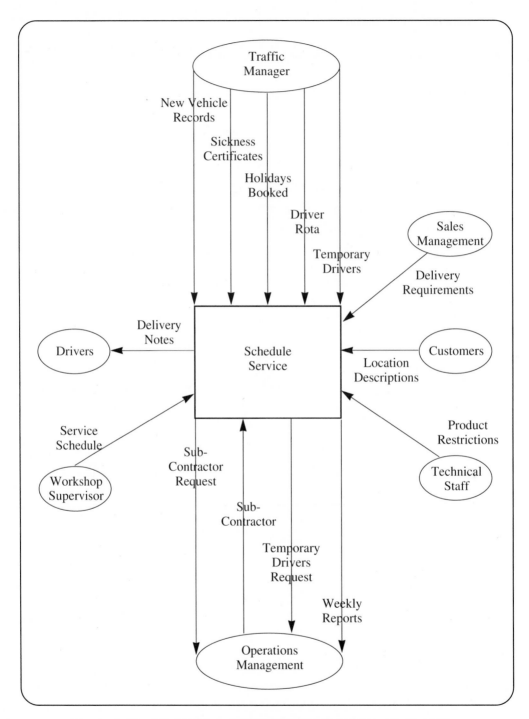

**Figure 3.17   Alternative context diagram for delivery scheduling
(Derived from the document flow diagram: 3.16)**

3.8.2   Function Definition

The more detailed definition of processes and the consideration of events will inevitably highlight shortcomings, misunderstandings and omissions in the required logical DFD. Thus, these subsequent modelling stages will feed back into the DFM and relevant changes will be made to ensure that the DFM is up to date.

3.8.3   Data Catalogue

The Input/Output Descriptions of the DFM will contain data items which in turn will be entered as Data Item Descriptions in the Data Catalogue. An example is given in Figure 3.20.

3.8.4   Requirements Catalogue

Many requirements are reports which give rise to retrieve only processes. These may be shown on the Data Flow Diagram if they can be conveniently grouped by the stores they access or the frequency with which they occur. However, an entry in a Requirements Catalogue (Figure 3.21) may be more appropriate.

3.9   SUMMARY

The Data Flow Diagram is the major modelling technique for describing the operations of the current system. It has three significant features:

Graphical, hence applying the dictum that a picture is worth a thousand words;

Consistent and easy to use hierarchy, which allows diagrams to be presented at appropriate levels of detail;

Process orientated, which makes them extremely accessible to users who often perceive systems in terms of processes or transformations. Properly used, DFDs are extremely effective models for communicating with system clients, operators and users.

The current and proposed logical Data Flow Diagrams retain the same notation and hence the advantages listed above. Furthermore, the notation introduced within a familiar user framework (the current physical system) should be assimilated by the advent of logical analysis and design. Consequently, new ideas and terms are introduced within a consistent notation.

Data Flow Diagrams (DFDs) are a fundamental modelling technique of most structured methodologies. They were introduced by de Marco (1979) and Gane and Sarson (1979) in the two seminal books that laid the foundation of the structured systems movement. They have also been extended to real-time development (Ward and Mellor, 1986) and generally they form the backbone of most proprietory methodologies.

DFDs have always been a fundamental model of SSADM and its proprietary cousin LSDM. Version 4 has clarified most of the outstanding notation and construction issues as well as giving some useful extensions.

However, the issue of bottom level process description requires further consideration.

SSADM has always been non-prescriptive in the way that the bottom level process is defined on an Elementary Process Definition. This loose approach is justified by stating that the detail of the process will be expressed in the dedicated process modelling techniques. However, it is difficult to subscribe to such looseness and we would prefer to encourage rigour as early as possible in process definition. Thus Elementary Process Definition should use precise methods such as Structured English, Decision Tables or Action Diagrams. These are described in detail elsewhere (*Skidmore and Wroe*, 1988, 1990).

Finally, DFDs are extremely well supported by CASE tools which enforce the completeness of the models and control consistency between the different hierarchical levels.

External Entity Description						*SSADM*
Variant:						Version 4
Project/System		Author	Date	Version	Status	Page of

ID	Name	Description
	Traffic Manager	Reports to the Managing Director Reported to by Operations Clerks Operations Supervisor Current occupant :- Jonathan White Overall responsibility for the operational running of the company. Hires drivers, sub-contractors and maintains holiday records.
	Drivers	Vehicle or Unit drivers Report to Operations Supervisor Currently 36 full-time and 4 part-time drivers

© National Computing Centre Limited, 1991   SSADMv4 21

Figure 3.18   External Entity Description

I/O Structure Description					*SSFDM* Version 4
Project/System	**Author**	**Date**	**Version**	**Status**	**Page of**

Data flows represented

I/O structure element	Data item	Comments
Driver	Driver number	
	Driver name	
Holiday day bookings	Days booked- hol	⎫
	Start date - hol	⎬ Repeating Group
	End date - hol	⎭
Holiday time bookings	Time booked- hol	⎫
	Date- time- hol	⎬ Repeating Group
	Start time - hol	
	End time- hol	⎭
Book date	Date booked- hol	

SSADMv4 27

Figure 3.19   I/O Structure Description for the Holidays Booked data flow of the delivery scheduling system

Attribute/Data Item Description						SSADM Version 4

Project/System EXAMPLE	Author SRS	Date 11/1	Version ID	Status D	Page 1 of 1

Attribute/data item name **Start Date-hol**	Attribute/data item ID
Cross-reference name/ID	Cross-reference type

End Date-hol	Attribute/Data Item
Days Booked-hol	Attribute/Data Item

Synonym(s)

Description

Start date of a holiday booking

Validation/derivation

Any valid work-date

Mandatory ☐ ✓ Default value **NONE**	Optional ☐ Value for null
Logical format **DO / MM / YY**	Unit of Measure **N/A**
Logical length **8**	Length description **N/A**

User role	Access rights
Traffic Manager	R/W
Traffic Staff	R only

Owner Traffic Manager

Standard messages **N/A**

Notes

Start Date-hol must have End Date-hol
Only used for full day bookings

SSADMv4 1

Figure 3.20   Attribute/Data Item Description

Requirements Catalogue Entry					SSADM Version 4

Project/System EXAMPLE	Author SRS	Date 11/1	Version 1C	Status D	Page 1 of 1

Source Traffic Manager	Priority 1	Owner Traffic Dept	Requirement ID R6

Functional requirement

Temporary Driver Usage Report/Enquiry

Non-functional requirement(s)

Description	Target value	Acceptable range	Comments
	Not applicable		

Benefits

Monitoring of temporary driver use.
Better management of full time staff

Comments/suggested solutions

Report layout (R107) produced and agreed

Related documents

R107 : Report layout : Temporary Driver Usage report

Related requirements

R109 : Sub-contractor usage report

Resolution

Report agreed

SSADMv4 42

Figure 3.21   Requirements Catalogue Entry

4 Logical Data Model

4.1 INTRODUCTION

A Logical Data Model (LDM) is a representation of the data used by a system. It shows how the data is logically grouped and the relationships between these logical data groupings as defined by the business requirements of the system.

The Logical Data Model comprises:

- a diagram – the Logical Data Structure (LDS);

- associated documentation of entities, relationships and attributes.

The LDS is initially developed to show the logical structure of the data used in the current system. This model is then refined, and may be extended, to show the logical structure of the data for the required system.

The required system LDS is subsequently used as basic information for the construction of the Entity Life Histories following the selection of a Business System Option. It is also used as a basis for physical data design.

4.2 MODELLING NOTATION

The model can be drawn using only two basic constructs, entity and relationship.

4.2.1 Entity

An entity is a logical grouping of data which is relevant to the application under investigation. It is an identifiable object, concept or activity belonging to the application about which descriptive data is stored.

Strictly speaking, there is a distinction between the terms *entity* and *entity type*. An entity type is the definition of all entity occurrences of the same type; its physical equivalent may be a file of records. An entity is an occurrence of an entity type which can be uniquely identified by some data values; its physical equivalent is a record occurrence. Normally the distinction between an entity type and an entity or entity occurrence is understood from the context.

Entities are represented on an LDS by a soft box. Each entity has a meaningful name, normally a noun, which is always singular. The name is placed inside the soft box (see Figure 4.1).

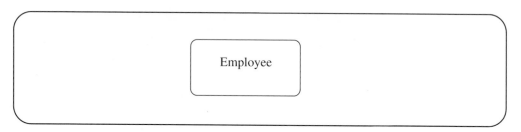

Figure 4.1   An example entity

In order to distinguish one entity occurrence from another, each entity will have defined an attribute, or group of attributes, whose value can uniquely distinguish between one entity occurrence and another. This important attribute(s) is called the identifier or key of the entity. The term attribute is used in preference to data item, field or data element in order to emphasise the logical nature of the LDM.

The entity Employee could use as its identifier the attributes employee name and date of birth, or a designed code such as employee number, as long as each occurrence can be uniquely identified.

4.2.2   Relationships

A relationship is a relevant business connection between two entities. A relationship is represented on an LDS by a line linking the associated entities.

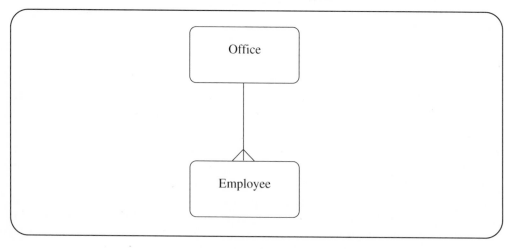

Figure 4.2   Example relationship between two entities

In SSADM, relationships are normally of degree one to many (1:m). For example, an Employee is allocated to exactly one Office, but each Office must have one or more Employees allocated to it. We express this by saying that there is a relationship of one to many between an Office and an Employee. The 'many' aspect of a relationship is represented by the 'crow's foot' symbol on the end of the relationship line (Figure 4.2).

A 1:m relationship defines the association between a master entity and a detail entity. From each master occurrence, it should be possible to access all detail entity occurrences, so for each office occurrence it is possible to access the group of employees who have been allocated to that office. From each detail entity occurrence, it should be possible to access the master occurrence, that is, for each employee it is possible to access the one office to which that employee has been allocated.

However, it is possible that an initial consideration of associations between entities reveals other types of relationships.

4.2.3   One to One Relationships (1:1)

If each office was allocated only one employee and each employee allocated to only one office, then the relationship between Office and Employee would be one to one.

As a general rule, one to one relationships are discouraged in SSADM and it is normal to merge the two entities involved into one entity. This may be reasonable if the entities have the same identifier. For example, a company may hold information about its employees (such as date of birth, sex, address) and hence an Employee entity is suggested. Furthermore, the sales details for each employee who is a salesman (such as number of orders, value of orders, sales territory, etc.) may have been logically grouped into a Sales entity and also given the identifier employee number. In this case it may be feasible to merge the Sales Entity and the Employee entity.

However, there are dangers in merging two entities involved in 1:1 relationships which do not have the same identifier. In the case of the one-office-to-one-employee relationship, merging the entities raises the question of what the identifier will be. If the identifier is employee number, then when the employee leaves the company it is likely that the entity occurrence will be deleted, with all its associated attributes. This means that if the office is not reallocated prior to the deletion then the attribute values of the office (office number, floor number, square footage, telephone number, etc.) will be lost. However, if the office number is made the identifier, it means that details of new employees cannot be recorded until they have been allocated an office that is free.

If the entities of 1:1 relationships are not to be merged then careful thought must be given to the question of the master and detail structure. It would be normal to make the entity whose occurrences were recorded earlier in time, or were the more permanent within the structure, the master; the entity which occurred second or was more transient in nature would become the detail. For example, even though an office only ever had one occupant, it would be more prudent to make office the master entity since it is probably more permanent in the business than the employees who may come and go. Thus a 1:1 relationship can be viewed as a special case of a 1:m relationship.

4.2.4   Many to Many Relationships (n:m)

Many to many (n:m) relationships occur quite regularly. For example, an employee may be contracted to work on many projects and a project may have many employees contracted to it (Figure 4.3).

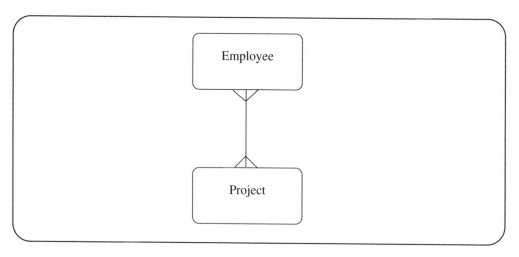

Figure 4.3   Example of a many to many (n:m) relationship

However, this type of relationship destroys the concept of master and detail and makes access more difficult; consequently, it is discouraged in SSADM.

Thus n:m relationships are decomposed into two 1:m relationships with the definition of an additional link entity (Figure 4.4).

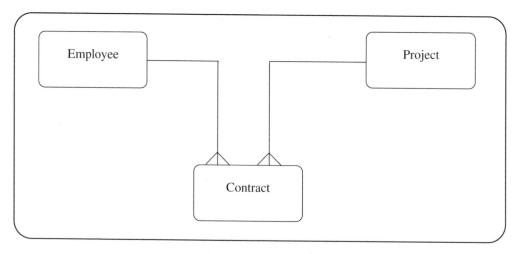

Figure 4.4   Decomposition of a n:m relationship

This allows all the contract details with which an individual employee is associated to be accessed as detail entity occurrences. It also allows access from the project entity of all the contracts associated with a specific project occurrence. Attributes of the original relationship can now be recorded as attributes of the link entity, eg date employee contracted to the project, duration of each contract. Note that the name of the link entity is normally the noun form of the verb which described the relationship; thus, the contracted relationship is replaced by a Contract entity.

4.2.5   Relationship Naming

The nature of the relationship between two entities is clarified by relationship naming and identification. A relationship link phrase is constructed from the perspective of each entity. For example:

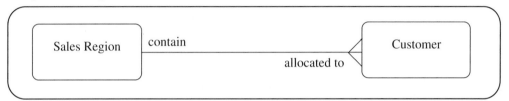

This reads from the Sales Region end as:

each Sales Region must contain one or more Customers;

and from the Customer end as:

each Customer must be allocated to one and only one Sales Region.

Each master entity may have a relationship with any number of detail entities. Likewise, an entity may be the detail of more than one master entity. It is also possible to have more than one relationship between two entities. In these circumstances the naming of relationships becomes particularly important (Figure 4.5).

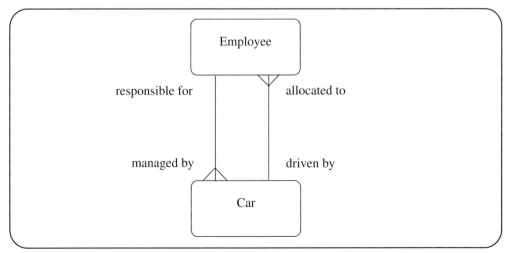

Figure 4.5   Naming of relations between two entities

each Employee must be responsible for one or more Cars;

each Car must be managed by one and only one Employee;

each Car must be driven by one or more Employees;

each Employee must be allocated to one and only one Car.

4.2.6   Optional Relationships

The example given in Figure 4.5 illustrates the careful specification of the obligatory nature of the relationship, must be . . . , one and only one . . . etc. This raises the issue of relationships being optional or mandatory.

A relationship is mandatory if an entity occurrence cannot exist without taking part in the relationship. For example, each occurrence of Employee must be responsible for at least one Car. Thus each Employee occurrence is associated with at least one occurrence of Car. If this is not true, eg Employee occurrence 31234 is not responsible for any cars, then the relationship is optional. A dashed line is used to represent an optional relationship. It shows that the entity occurrence at that end of the relationship need not be associated with entity occurrences at the other end.

This allows the more detailed definition of relationship types. Let us return to examples used earlier in the chapter.

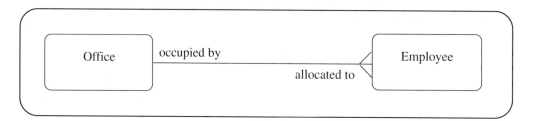

each Office must be occupied by one or more Employees;

each Employee must be allocated to one and only one Office.

However, changing the nature of the relationship leads to the following possibilities

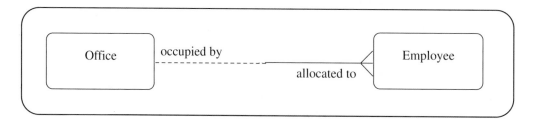

each Office may be occupied by one or more Employees;

each Employee may be allocated to one and only one Office;

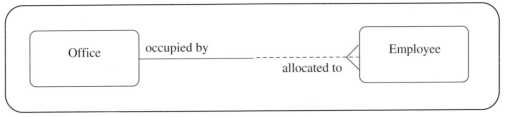

each Office must be occupied by one or more Employees;

each Employee may be allocated to one and only one Office;

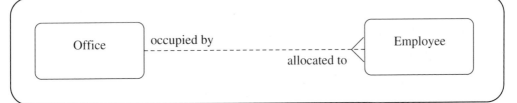

each Office may be occupied by one or more Employees;

each Employee may be allocated to one and only one Office.

These examples give a clue to the format of the relationship statement

        each

        Entity Name

        must be / may be

        link phrase

        one and only one / one or more

        Entity Name

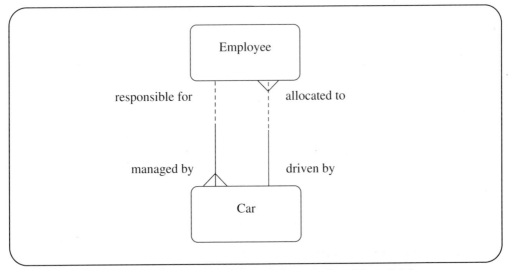

Figure 4.6   A re-drafting of the relationships of 4.5

So, redrawing Figure 4.5 (Figure 4.6) gives:

each Employee may be responsible for one or more Cars;

each Car must be managed by one and only one Employee;

each Car must be driven by one or more Employees;

each Employee may be allocated to one and only one Car.

4.3   CONSTRUCTING THE LOGICAL DATA MODEL

The basic steps in LDM construction are summarised below:

- develop initial entities;
- identify initial relationships;
- validate the LDS against the current system;
- extend the LDS to cover requirements;
- rationalise the structure;
- revalidate the LDS.

1. Develop initial entities

The initial entities will be the objects about which the system currently holds data. These will be identified from the interview notes, documents and observations made in the fact-finding and investigation of the current system.

Each entity occurrence must be uniquely identifiable and there will normally be more than one occurrence of an entity. Existing file content and searching and retrieval procedures give vital clues to entities of the system.

Figure 4.7 shows a Data Flow Diagram of a simplified library loans and reservation system.

The following processes have been identified:

- Store Borrower details;
- Record acquired Book-Copies;
- Make a Loan;
- Record the return of a Loan;
- Record a Reservation;
- Delete a Book;
- Delete acquired Book-Copies;
- Compile Loan History Report;
- Compile Reservation History Report.

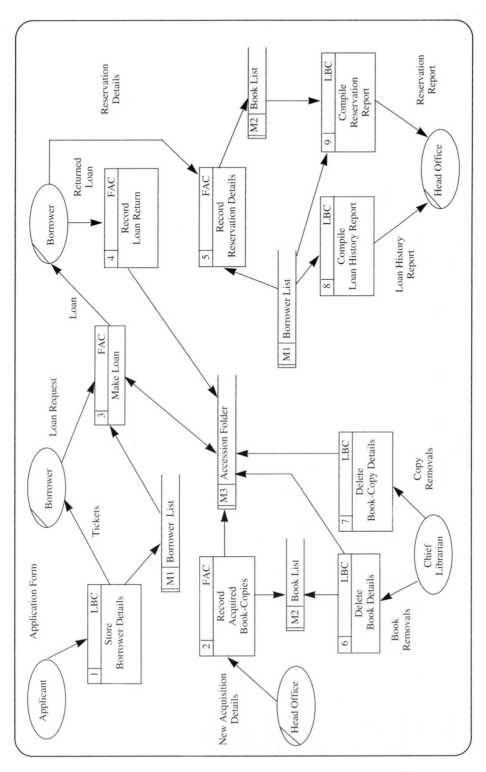

Figure 4.7 Simplified library loans and reservation system

The processes and stores of this current physical DFD (Figure 4.7) suggest the following initial entities

Borrower	identified by	membership number
Book	identified by	ISBN
Book-Copy	identified by	accession number

Loan and Reservation are also possible entities. However, there is some feeling that on-loan and reserved may only be attributes of Book and Book-Copy and so they are omitted at this stage.

2. Identify initial relationships

Successful identification of relationships between entities demands constant reference to the business requirements and environment of the system. Consider the following entities for a simple order processing system

Order	identified by	order number
Order Line	identified by	order number, product code
Product	identified by	product code
Supplier	identified by	supplier reference number

The two following data structures support different business requirements (Figures 4.8 and 4.9).

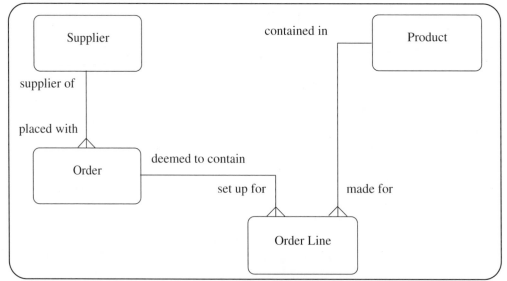

Figure 4.8   Order processing LDS with Supplier and Product related through Order and Order Line

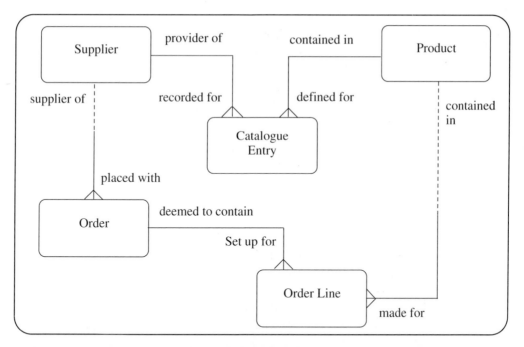

Figure 4.9   Order processing LDS with Product and Supplier related through Catalogue Entry

The relationships for Figure 4.8 are as follows:

each Supplier must be supplier of one or more Orders;

each Order must be placed with one and only one Supplier;

each Product must be contained in one or more Order Lines;

each Order Line must be made for one and only one Product;

each Order must be deemed to contain one or more Order Lines;

each Order Line must be set up for one and only one Order.

In the model given in Figure 4.8 product details are only held if an order (and hence an order line) has been raised for this product. Similarly, supplier details are only stored if an order has been placed with that supplier.

However, if the business rules require the model to show all products that can be ordered from a supplier, irrespective of whether they have ever been ordered, then the relationships change quite dramatically (Figure 4.9).

The relationships of Supplier with Order and Product with Order Line become optional at the Supplier and Product end. Thus supplier entity occurrences may now be held which are not associated with an order. Similarly, product details may be held for products which have never been ordered.

The new entity Catalogue Entry is created out of the decomposition of the n:m relationship

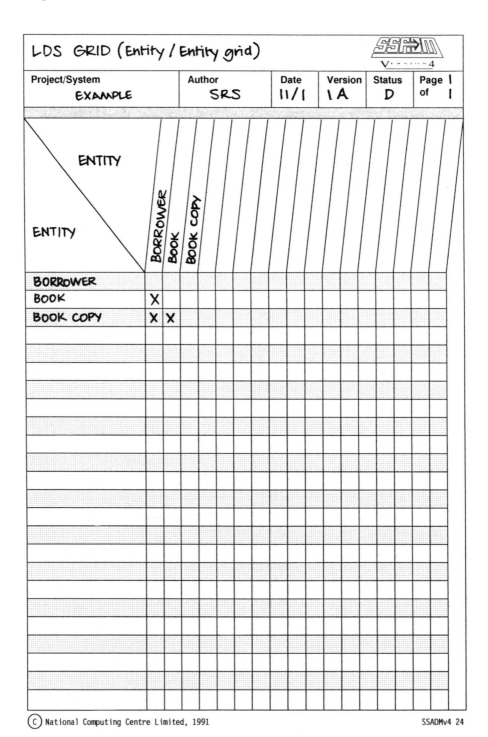

Figure 4.10   LDS Grid for library loans

between supplier and product. The identifier of Catalogue Entry is supplier reference number, product code and the entity will probably contain such data items as price and delivery terms. The relationships are mandatory:

each Supplier must be provider of one or more Catalogue Entries;

each Catalogue Entry must be recorded for one and only one Supplier;

each Product must be contained in one or more Catalogue Entries;

each Catalogue Entry must be defined for one and only one Product.

The important point to make is that both of these models are 'correct' given different business requirements.

Some analysts find that relationships may be clarified through the construction of an entity/entity grid. This is built up to show which pairs of entities are directly related to each other. Entities are directly related if the relationship between them can be described without involving other entities. For example, *a borrower loans a book copy* is a direct relationship.

Directly related pairs of entities are indicated by an 'X' in the cell of the matrix at the intersection of the relevant row and column. Only half the grid actually needs completing.

Figure 4.10 shows the grid of relationships. An LDS can be built from the entities and relationships shown on the grid. It is necessary to decide whether the relationships are 1:1, 1:m or n:m. For 1:1 relationships, a decision has to be made whether to merge the two entities associated with the relationship or to make one the master and one the detail. The n:m relationships need breaking down using a link entity.

Figure 4.11 shows the LDS resulting from the grid in Figure 4.10. Note the emergence of Loan and Reservation as link entities. Thus if these had not been established as initial entities they will now emerge from decomposition of the n:m relationships.

Figure 4.11 also reflects a number of enterprise rules that have been uncovered in fact finding interviews:

- Details of a Borrower may need to be recorded and retained even if the borrower is not associated with a Loan.

- Details of a Book are not recorded until at least one copy of the book is available.

- However, it is possible for a Book Copy to never participate in a Loan. If the relationship between Book Copy and Loan is made mandatory at the Book Copy end, then an entity occurrence for Book Copy could not be created until it is loaned out for the first time. This is clearly unsatisfactory.

- Similarly a Book may never be reserved and a Borrower never make a Reservation.

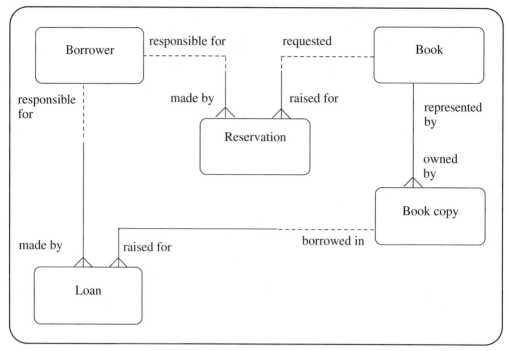

Figure 4.11   LDS for the simplified library system

The following relationships are supported in Figure 4.11:

each Borrower may be responsible for one or more Loans;

each Loan must be made by one and only one Borrower;

each Book Copy may be borrowed in one or more Loans;

each Loan must be raised for one and only one Book Copy;

each Borrower may be responsible for one or more Reservations;

each Reservation must be made by one and only one Borrower;

each Book may be requested in one or more Reservations;

each Reservation must be raised for one and only one Book Copy;

each Book must be represented by one or more Book Copies;

each Book Copy must be owned by one and only one Book.

3. Validate the LDS against the current system

The initial LDS is drawn to represent the scope of the current system and validated with users. The word *scope* is very important here. The LDS is logical and so does not show physical aspects of the current system, hence entities such as Blue Enquiry Form, Employee Master Listing and Order Form are not permissible. It is their logical equivalents (Enquiry, Employee and Order) which will be shown on the LDS. However, at this

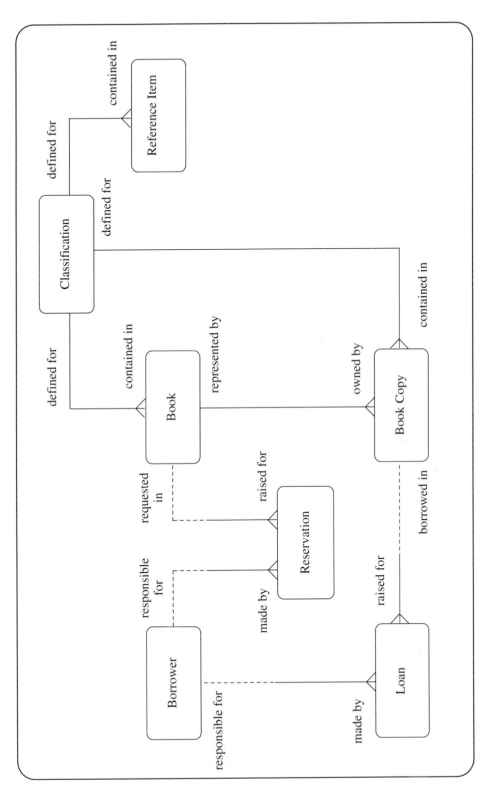

Figure 4.12   Required LDS for the simplified library system

stage, the LDS only represents entities identified within current operations and not those associated with requirements. This current LDS can be cross-referenced to the current logical DFD (see section 4.7).

4. Extend the LDS to cover requirements

The LDS is now extended to cover entities needed to fulfil the new requirements of the system.

In the library example there is a need to provide a catalogue of all books and reference items. This introduces two further entities:

Classification identified by classification code.

Reference Item identified by reference code.

These are added to the LDS (Figure 4.12) and the entity/ data store cross-reference extended to include the entities of the required LDS and the data stores of the required logical DFD.

It must be recognised that data modelling is an iterative process. An LDS is constructed and repeatedly tested against the business requirements of the system. This iterative process includes revalidation after each modification within a stage and at other stages within the methodology as more information about the data structure emerges.

Validation concerns not only whether the LDS meets the data requirements of the system, but also whether it supports the processing requirements as well. Each process will need to navigate the data structure by means of the access paths provided by the relationships. As each requirement, and the processing to meet it, is determined the LDS should be examined to ensure that the paths are adequate and will provide the correct processing or answers to queries.

For example, the query 'Which borrowers currently have on loan a copy of a specified book?' can be answered by using the ISBN as the entry point to Book and then following the master to detail relationship to Book Copy and similarly to Loan. For each current loan, the detail to master relationship will access the Borrower information. These access paths will be modelled more formally in Enquiry Access Paths (see Chapter 8).

5. Rationalise the structure

A review of the LDS is required to ensure that it is the minimum logical model required to support the system requirements.

In the library example two constructs require consideration.

The first is where there are two access paths between two entities such that access logic is duplicated and perhaps redundant. Figure 4.12 shows a triangle of relationships between Classification, Book and Book Copy indicating that there is duplicate logic between Classification and Book Copy. It may be possible to remove one of these relationships. The prime candidate for removal would be the relationship giving the shortest route, that is, the relationship between Classification and Book Copy. The classification of a book copy could still be found by accessing the relevant Book occurrence for a Book Copy and hence the relevant Classification occurrence.

A problem may also arise where link entities have been created in the resolution of n:m relationships. If these are found to contain similar data, it may be that they can be merged. For example, consideration might be given to merging Reservation and Loan. This issue is considered later.

4.4 FURTHER NOTATION AND CONSTRUCTS

A valid LDS can be produced using the basic notation and constructs described above. However the notation, and hence information content, of a LDS can be enriched by a few simple extensions to this basic notation.

4.4.1 Exclusive Relationships

In an exclusive relationship the participation of an entity occurrence in one relationship precludes it from participating in another. For example, in Figure 4.13 an Employee may work from an Office or a Sales Territory but not both. This is indicated by an exclusivity arc.

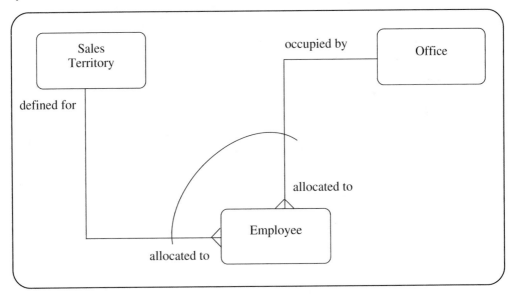

Figure 4.13   Example of an exclusive relationship

    each Employee must be allocated to one and only one Sales Territory OR to one and only one Office;

    each Office must be occupied by one or more Employees;

    each Sales Territory must be defined for one or more Employees.

Similarly, a master occurrence may have a detail occurrence which is one of a number of options (Figure 4.14).

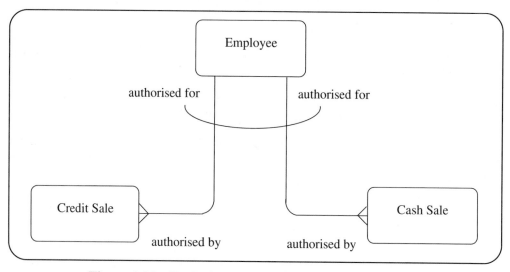

Figure 4.14   Exclusive relationships for a master entity

In Figure 4.14 an employee deals either with cash sales or credit sales but not with both.

The exclusive relationship may extend to more than two alternatives.

4.4.2   Recursive Relationships

There are situations in which entities have a relationship with themselves. For example, each employee is assigned a supervisor who is also an employee. Each employee who is a supervisor may supervise many other employees. This example of a recursive relationship is often drawn as in Figure 4.15.

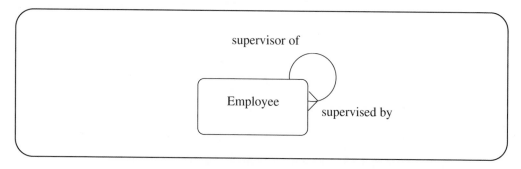

Figure 4.15   An example of a recursive relationship

However, there are employees who are not supervised and other employees who are not supervisors. Thus, the recursive relationship is optional as in Figure 4.16.

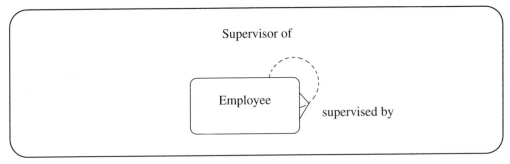

Figure 4.16   An optional recursive relationship

Recursive relationships may also be used to represent a natural hierarchy of master and detail.

For example, an organisation hierarchy can be modelled as a master detail hierarchy or may be represented by a single entity called Organisational Unit with a recursive relationship which defines which unit oversees other units (Figure 4.17).

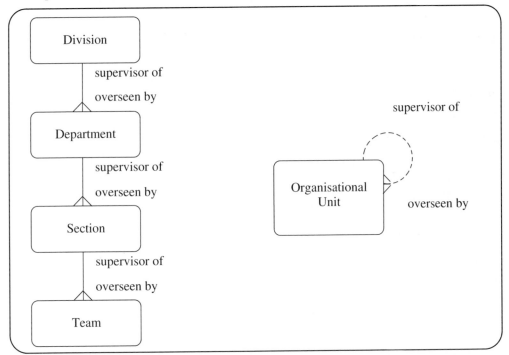

Figure 4.17   Alternative structures for representing an organisation hierarchy

There are occasions when a recursive relationship is n:m. For example, in a bill of materials a part may be a component of many other parts, while it may itself be made up of many other parts. To resolve the n:m relationship, a link entity Component is defined with two 1:m relationships with Part (Figure 4.18).

Thus the Components of each part will form the detail of Part, and it will also be possible to find of which Components a Part is a constituent.

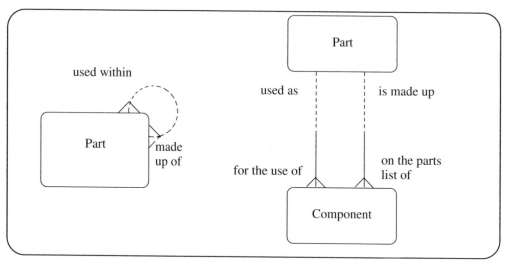

Figure 4.18   Resolution of a recursive n:m relationship

4. 5   VALIDATION OF THE LDS

The danger with rationalisation (see section 4.3) is that in removing what appears to be duplication, some information may be lost. Consequently, it is necessary to revalidate to ensure that the LDS still satisfies all the requirements. For example, returning to the simplified library system of section 4.3, the duplicate access path between Classification and Book Copy was considered redundant. However, further investigation has confirmed that a book can belong to a number of classifications. Consequently, it is the practice to assign one or more copies of a book to each of the relevant classifications. This ensures that a physical copy of a book will be shelved in the physical library area of each classification. Hence it is the relationship between Classification and Book which is not only redundant but wrong in this instance.

Continued investigation reveals that it is not possible to merge Loan and Reservation as it is confirmed that it is a Book Copy that is the subject of a Loan and a Book that is the subject of a Reservation.

Figure 4.19 shows the final model that is consistent with the requirements as revealed up to this stage.

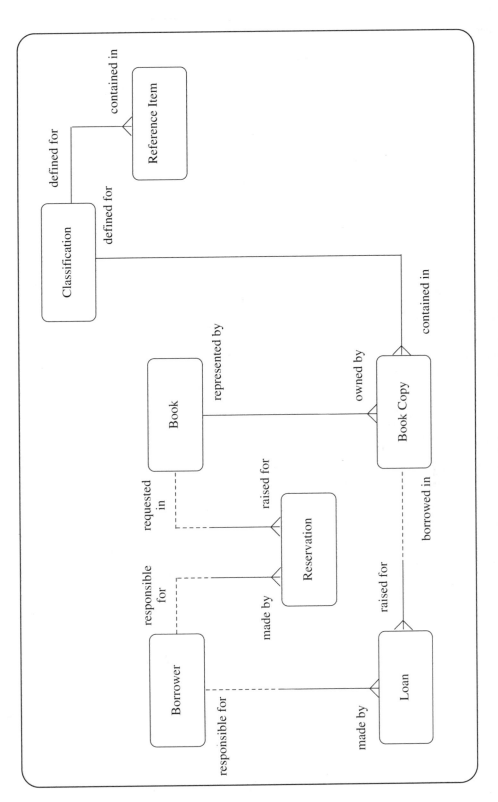

Figure 4.19 Revised required LDS for the simplified library system

4.6   DOCUMENTATION

The documentation of the Logical Data Model will include:

- The Logical Data Structure;
- Supporting entity descriptions. These add volume data to the entity descriptions and list the attributes associated with that entity. (See Figure 4.20.)
- Supporting attribute descriptions. Each attribute is defined within the Data Catalogue. (See Figure 3.20.)
- Relationship descriptions. (See Figure 4.21.)

Entity Description - Part 1						SSADM	
Variant: Required						V...4	

Project/System Example	Author SRS	Date 11/1	Version 1A	Status D	Page 1 of 1

Entity name BORROWER	Entity ID E1

Location N/A	Occurrences 5.000	Average 7.500	Max

Description

A person granted lending rights to the library

Synonym(s)

Attribute name/ID	Primary key	Foreign key
Member - no	✓	
Member - name		
Member - address		
Join date		
Member - type		
Loan - limit		

Rel no	'must be'/ 'may be'	'either'/ 'or'	Link phrase	'one and only one'/ 'one or more'	Object entity name
1	may be		responsible for	one or more	loan
2	may be		responsible for	one or more	reservation

Notes

SSADMv4 17

Figure 4.20a   Entity Description for Borrower

Entity Description - Part 2					SSADM
Variant: **Required**					Version 4

Project/System		Author	Date	Version	Status	Page I
EXAMPLE		SRS	11/1	I A	D	of I

Entity name BORROWER	Entity ID

User role	Access rights
Librarian Chief Librarian	Insert, Read, Modify Read, Relate, Archive

Owner   Chief Librarian

Growth per period

300 / year

Additional relationships

N/A

Archive and destruction

Archived one year after Borrower left library
Destroyed five years   "   "   "   "

Security measures

Data Protection Legislation under investigation

State indicator values

Not yet defined

Notes

SSADMv4 18

Figure 4.20b   Entity Description for Borrower (cont.)

Relationship Description					⠀SSADM⠀
Variant: Required					Version 4

Project/System EXAMPLE	Author SRS	Date 11/1	Version 1A	Status of D	Page 1

Entity name **BORROWER**   Entity ID **E1**

Mandatory ☐	Optional ☑	%Optional **2**

Link phase **responsible for**

Description **Links all the loans that a borrower has been and is currently responsible for**

Synonmyn(s)

Object entity name **LOAN**   Object entity ID **E2**

One (1:) ☐	Many (m:) ☑	Minimum **O**	Average **40**	Maximum **100**

Cardinality description **18% 1   60% 40   20% 100**

Growth per period   **10% per annum**

Additional properties   **N/A**

User role	Access rights
Librarian	**ALL**

Owner **Chief Librarian**

Notes

None

Figure 4.21   Relationship Description for Borrower may be responsible for Loan Relationship

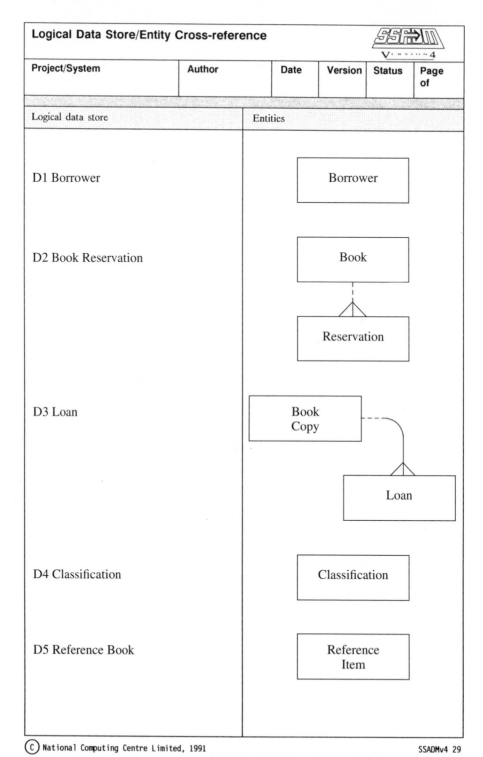

Logical Data Store/Entity Cross-reference

Project/System	Author	Date	Version	Status	Page of

Logical data store	Entities
D1 Borrower	Borrower
D2 Book Reservation	Book / Reservation
D3 Loan	Book Copy / Loan
D4 Classification	Classification
D5 Reference Book	Reference Item

SSADMv4 29

Figure 4.22   Example of data store/Entity cross-reference grid

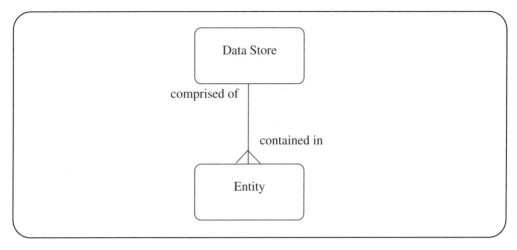

Figure 4.23   Relationship between stores and entities

4.7   RELATIONSHIP WITH OTHER MODELS

4.7.1   Data Flow Model

The Logical Data Model is developed simultaneously with the Data Flow Model, and hence it is important to ensure that the two are consistent. This is formally achieved by cross-referencing the stored data of the DFM (data stores on the DFD) with the stored data of the LDM (entities and their supporting descriptions). The data items of the DFM and the attributes of the LDM are, of course, the same thing.

A data store/entity cross reference documents this relationship (see Figure 4.22). This chart shows which entities are in which data store. The relationship between these two constructs is given in Figure 4.23.

each Entity must be contained in one and only one Data Store;

each Data Store must be comprised of one or more Entities.

4.7.2   Function Definition

Each update and enquiry function requires an access path through the LDS to obtain the necessary data. The update functions will largely be defined on the Effect Correspondence Diagram (ECD) and Enquiry Access Paths (EAP). The construction of these will feed new information back into the LDM.

4.7.3   Entity-Event Modelling

Each entity will be modelled on an entity life history (ELH). These may reveal new attributes, new entities and new relationships which have to be reflected in the LDM.

4.7.4   Relational Data Analysis

Relational Data Analysis (RDA) is complementary to Logical Data Modelling. RDA is applied to I/O structures and the results are used to validate the required LDM. A composite data design will emerge as a result of combining the 'best of' the top-down and bottom-up approaches.

4.8   SUMMARY

Logical Data Modelling is a major modelling method within SSADM which enables a logical view of the structure of the data in the system to be developed. It is a graphical technique which uses the two basic constructs of entity and relationship, although these can be enriched by additional constructs.

Logical Data Models are found in most structured techniques. Alternative names for this model include entity model, data entity model, entity-relationship diagram and ERA model (Chen 1976, Howe 1983, Benyon 1990, Martin and McClure 1985).

SSADM Version 4 provides four improvements over the previous release:

- the ability to explicitly show optionality at both ends of the relationship;

- a more definitive, comprehensive and standard method of relationship naming;

- more comprehensive supporting documentation in the LDM;

- the dropping of Operational Masters which introduced physical access issues too early in the logical modelling cycle.

It is important that the Logical Data Model and the Data Flow Model are consistent with each other and are kept consistent throughout systems development. Experience often suggests that the DFD is not always amended to reflect changes in the LDS and hence becomes an unreliable perspective of the system. This should be guarded against.

5 Functions, Dialogues and Prototyping

5.1 INTRODUCTION

This chapter focuses on the way that users interact with the system through the dialogues or conversations they use to enter inputs and request outputs. It must be recognised that this is a significant area of systems design because for most users the inputs, outputs and dialogues *are* the system and hence its overall utility is often judged from these factors alone. SSADM approaches dialogue definition through a consideration of the *functions* of the system and the *roles* of its users.

It also recognises that examination of, and experimentation with, different dialogue structures will help users further clarify their requirements. This definition is easier where the user and analyst can use a *prototype* system or part-system. Hence prototyping is encouraged and documented in SSADM and will, in turn, feed back into the logical models which have formed the basis of design so far.

The structure of this chapter is summarised in Figure 5.1.

5.2 FUNCTIONS

Functions are sets of system processing which the user wishes to carry out at the same time to support the business activity. A function consists of the input, the processes (both update and enquiry) required to respond to that input, and the output produced as a result. Functions are perhaps best described in context and hence examples are given in this section to provide typical circumstances. However, it is worth stating (and the point will be made again) that functions can only be described in the business and user context and therefore strict rules of definition cannot be formulated.

The functional perspective of the system has already been considered in the processes and Elementary Process Descriptions of the Data Flow Model (DFM). Further functional issues are considered in events which are examined in detail in the Event/Entity Matrix and Entity Life Histories (see Chapter 7). Briefly, an event is defined as something that triggers a process to update system data. An event is not a process, it is the stimulus which causes that process to be invoked.

Three types of event can be distinguished:

- external event   – a transaction arriving from the outside world;

   – internal process event – occurs when a predefined condition within the stored data has been met;

   – time-based event   – takes place when a particular process is to be triggered at a regular time interval: a set time of day, month, year, etc.

A single event instance may cause more than one entity occurrence to change. The changes within a single entity occurrence caused by an event is called an effect. Events and effect are considered in more detail in Chapter 7.

The relationship of functions to events and processes is worth detailed consideration.

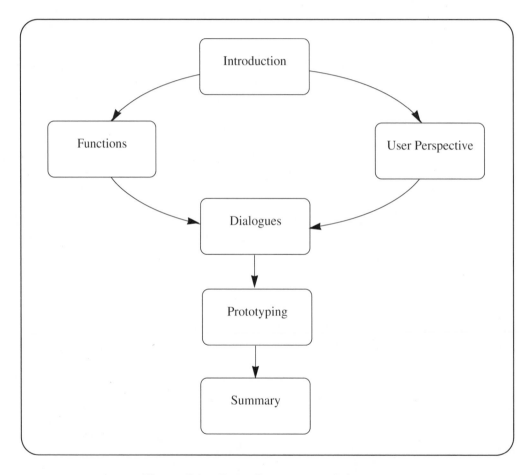

Figure 5.1   Overall structure of chapter

5.2.1   Relationship of Functions to the Processes of the DFM

The Data Flow Model (DFM) has defined the processes that should be supported in the required system. How each of these processes is related to a functional business activity can only be defined in consultation with the user.

Three possibilities exist:

1:1   Each bottom-level process on the Required System DFD becomes a function. Thus the process Create Order becomes the function Order Details created. This occurs when the user agrees that the process encompasses a well-defined business activity. The actions specified within the process are undertaken at a certain time and no part of the process can take place independently.

1:M Certain bottom level processes become many functions. For example, the bottom level process Create Order may be defined in a one page Elementary Process Description and hence fulfils the technical requirements of the DFM. However, on closer inspection of the activities within that process, the user may perceive that several business functions are covered by the definition. For example, it may contain the following:

Creation of new customer details. If the order is from a new customer then certain credit checks have to be run before the order is progressed. This may be done at a different time (removing such orders from the order handling flow whilst checking takes place) or at a different place (accounts sub-system).

Furthermore, the user may reveal that a new customer can be created before an order is placed and so this function needs to work independently. A failure to recognise this may lead to a system where a new customer can only be created when an order has been received from them.

Order Details created

Standing Order Details created. This may simply be an order type data item on the process definition. However, it may have organisational implications and be handled by a different department.

M:1 Many low level processes of the Required Systems DFM are combined into one function. Thus the processes of the DFD

Validate Order

Create Order Details

Print Pro-forma Invoice

are perceived by the user as one business function (Order details created) because none of these processes are performed independently and they are all undertaken by the same user at the one time. This overall *cohesiveness* is an important guide to function identification.

This combination of processes into one function is particularly common when the DFD developer has used inter-process flows on the diagram (see Figure 5.2). Thus function definition also attacks the problem of developers diagramming at different levels of detail.

The Required System DFD will provide important clues about the user initiated functions that have to be supported. Some enquiry functions may be shown on the DFD or will be available from the Requirements Catalogue. However, temporal and state driven functions will be derived from a consideration of the events which will eventually drive the construction of Entity Life Histories (ELH) – see Chapter 7.

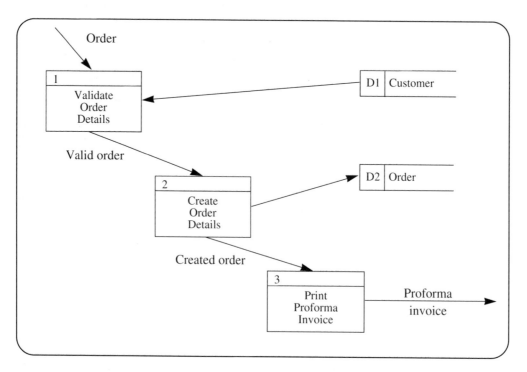

Figure 5.2   Part DFD showing three processes destined to be combined into the function order details created

5.2.2   Relationships of Functions to Events of the ELH

Similar relationships occur between functions and events.

1:1 An event 'Receipt of order details from existing customer' triggers the function order details created.

1:M An event 'Receipt of order details' triggers the functions order details validated and order details created.

M:1 Two events 'All of order cleared from quay' AND 'Two years since order landed' are needed to trigger the function order details deleted.

The function may also by triggered by alternative events. For example, order details are deleted when 'All of order cleared from quay' OR 'Two years since order landed'.

The events will assist the definition of system initiated functions and will need to be identified for the Event/Entity Matrix which precedes the construction of Entity Life Histories.

Thus the definition of a function and its relationship to processes and events must be considered:

- in the context of the user's business activity and the way that activity is conducted within the organisation. It must again be stressed that the user has a vital role to play in function definition.

- in the light of the variety of modelling detail produced by the developer. Function definition should even out problems of different DFD and event identification.

Two other aspects need highlighting:

- Function definition represents an important way of reconciling the two views of the system: processes from the DFM and events from the ELHs which in turn have come from the entities of the Logical Data Model (LDM).

- Function definition is iterative. The Function definition documentation (see Figure 5.4) is completed over time. Function definition may begin once the Required System DFM has been completed, with initial update functions defined from the processes of the DFD. More details about the function will be discovered during the

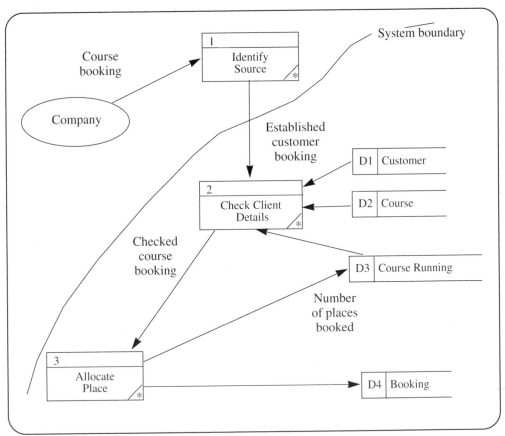

Figure 5.3   Part DFD for course booking system

construction of Entity Life Histories. The scope of functions may also be reviewed and additional functions defined. Hence the Function Definition is added to and amended as more is understood about the functional requirements of the system. It is a reference point for techniques, documents and activities and it can be perceived as the gradual building of processing blocks constructed from processes, events and enquiries. These functions will form the basis of physical design.

5.2.3   Function Definition: An Example

Figure 5.3 shows part of a Required Logical DFD for a course booking system. The functionality of processes 2 and 3 will be considered in detail.

Process 2

a Retrieves and displays customer name and address from the customer store.

b Retrieves and displays course name, course duration and course price from the course store.

c Retrieves and displays course run number, course run date, hotel, location, accommodation cost and places booked from the course running store.

Process 3

a Allocates the course bookings to particular courses.

   This;

b Creates an entry on the booking store. This will store booking number, course run number and delegate name.

c Increments the number of places booked on the course running store.

Discussion with the user will determine how many functions this covers.

A function for each process might be suggested. Hence one function retrieves and displays customer and course details and the second allocates a booking to a course. However, if the user usually wants to display details and then undertake the allocation, then a single function is more appropriate which is a combination of the two processes.

Further discussion with the user reveals that the retrieve and display function is often used to respond to customer and management queries and therefore does not have to be followed by the allocation process or preceded by customer details. Hence two functions are agreed.

1. Enquire on Course status. This function covers parts b and c of process 2.

2. Registration of course bookings. This function covers all of processes 2 and 3.

If the user always undertook function 2 after function 1 then the two functions could be combined. This is clearly not the case and so two Functional Definitions will be developed. The example for the registration of course bookings is given in Figure 5.4.

Function Definition						SSADM Version 4

Project/System Example	Author SRS	Date 10/1	Version	Status	Page 1 of 1

Function name **Registration of course bookings**	Function ID 1

Type **Update / on-line / user**

User roles **Booking Officer**

Function description

This function is invoked when bookings are received from an established customer. Client course and course running details are displayed and the bookings allocated to an appropriate course. A course booking entry is created as a result.

Error handling

Course and client details cannot be altered in this function

DFD processes **2, 3**

Receipt of booking form from established customers	Event frequency 10 / customer / year

I/O descriptions
I/O structures
Requirements Catalogue ref. **36 - 1**
Volumes **Average : 10 / day   Maximum : 75 / week**
Related functions **Course Status Enquiry**

Enquiries None	Enquiry frequency N/A

Common processing
Dialogue names

Service level requirements

Description	Target value	Range	Comments
On-line response time	2 secs	2 - 5 secs	None

Ⓒ National Computing Centre Limited, 1991   SSADMv4 22

Figure 5.4   Function Definition for Registration of Course Booking

5.2.4 Documenting a Function

The Function Definition (Figure 5.4) is progressively completed as information becomes available. Each function must be categorised as:

- an Enquiry or an Update. An update function may include enquiries;
- off-line or on-line. This depends upon whether the stored data of the system is updated or interrogated on-line or off-line;
- either User or System initiated.

5.2.5 Moving from Functions to Dialogues

The Required Logical Data Flow Diagram is supported by I/O descriptions for each data flow that crosses the system boundary (see Chapter 3). Established customer booking is the only I/O for the part course booking system introduced in the previous section. This is documented in Figure 5.5. The data items contained in the system responses to the input are not described on this form.

5.2.6 I/O Structures

The I/O Descriptions are converted into I/O Structure Diagrams using the notation of Chapter 2 (Figure 5.6(a)). The Structure Diagrams add the data items of the system response and are supported by I/O Structure Descriptions (see Figure 5.6(b)). The bottom leaves of the structure represent one or more data items crossing the system boundary and each is annotated as either an input or an output. Within the I/O structure:

- repeating groups of data items are represented as an iteration;
- optional groups of data items are represented by a selection with a null option;
- mutually exclusive groups are represented as options under a selection.

I/O Structures and their supporting descriptions are also constructed for enquiries.

5.2.7 Functions: Summary

Function definition identifies how to best organise the system processing to support the user's business tasks. In doing so, it reconciles the two views of processes derived from the process analysis model (the DFM) and the data orientated perspective (entity-event modelling). It defines constituent processes and events in terms which are cohesive in the context of the user's business activity. For each function an I/O Structure and its supporting description is constructed and these will form the basis of subsequent dialogue design.

5.3 THE USER PERSPECTIVE

The term 'user' is generally employed throughout the industry to describe people who

I/O Descriptions						SSADM
Variant: REQUIRED LOGICAL						Version 4

Project/System EXAMPLE		Author SRS	Date 13 / 1	Version	Status	Page 1 of 1

From	To	Data flow name	Data content	Comments
1	2	Established Customer Booking	CUSTOMER CUSTOMER NAME CUSTOMER ADDRESS COURSE CODE COURSE RUN DATE DELEGATE NAME	} could be repeating group

© National Computing Centre Limited, 1991   SSADMv4 26

Figure 5.5   I/O Description for Established customer booking Data Flow

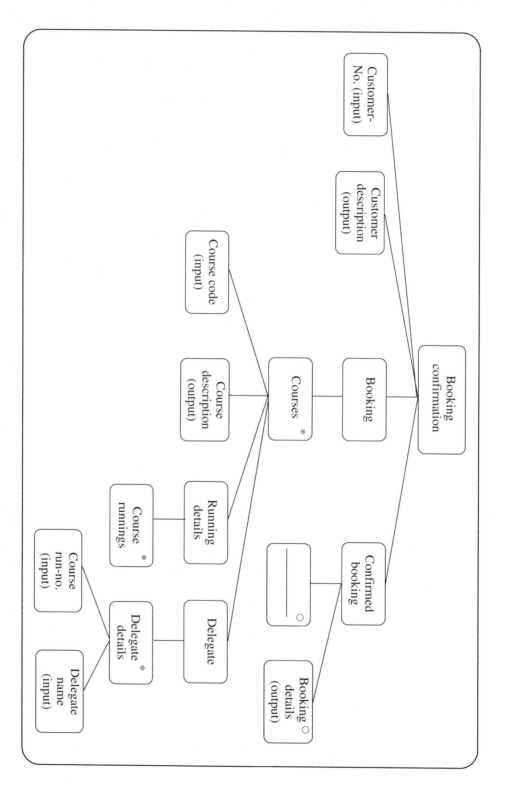

Figure 5.6a   I/O Structure Diagram for Registration of course bookings Function

I/O Structure Description					SSADM V........4	

Project/System EXAMPLE	Author SRS	Date 13/2	Version	Status	Page 1 of 1

Data flows represented **Established customer bookings**

I/O structure element	Data item	Comments
CUSTOMER NUMBER	CUSTOMER ※	Bookings only
CUSTOMER DESCRIPTION	CUSTOMER NAME	accepted from
	CUSTOMER ADDRESS	established customers
COURSE CODE	COURSE CODE	Maybe more than one
COURSE DESCRIPTION	COURSE NAME	
	COURSE DURATION	
	COURSE COST	
COURSE RUNNINGS	COURSE RUN ※	
	COURSE RUN DATE	
	LOCATION	} Repeating Group
	ACCOM-FEE	
	PLACES BOOKED	
COURSE RUN-NO	COURSE RUN ※	
DELEGATE NAMES	DELEGATE NAMES	Maybe more than one
BOOKING DETAILS	BOOK-DATE	System derived
	BOOK CLERK	Dialogue confirmed
	BOOKING-REF	System derived

SSADMv4 27

**Figure 5.6b   I/O Structure description for Registration of course bookings
Function**

will interact with the system in some way. However, this term will embrace at least the following:

- staff who use the system on a regular basis to enter data but do not produce or understand the relevance of various outputs. These might best be described as operators.

- regular users of output data who may have little direct interaction with the system but are reliant upon its output. In many respects these are the major users of the system.

- staff who require ad hoc information and use the system infrequently. This might include auditors and Data Protection staff.

- managers who can cause the termination of system development or remove the system by withdrawing their support. These are effectively the clients or customers for the system.

SSADM requires the definition of all users of the system and documents their responsibilities in a User Catalogue. This is used as a basis for establishing user roles. These roles may encompass more than one job title in circumstances where there is a significant overlap in responsibilities.

5.3.1   The User Catalogue

The job title and responsibilities of each user will be documented in a User Catalogue (see Figure 5.7). At the start of the project this will cover the operators and users of the current system. However, as development progresses the User Catalogue will begin to reflect the requirements of the replacement system. An organisation chart is a useful supplement to the catalogue. This will show how the jobs are formally related to each other in the enterprise.

5.3.2   User Roles

From the User Catalogue it may become clear that certain users largely undertake the same tasks. For example, in a course administration system the Sales Clerks and the Administrative Assistant both include 'recording course bookings' as a significant job activity in the User Catalogue. This common function defines a user role, 'Booking Officer', which is independent of the job title of the person who is actually making the booking. The purpose of defining this role is to group together common functions and so avoid specifying a separate dialogue for two different groups of user.

It must be stressed that a user role can only be defined for groups of users where there is a large overlap of common tasks. So, for example, the Sales Manager may also undertake course bookings between 5 pm and 6pm when the other booking staff have gone home. However, this only represents a very small proportion of his overall tasks, and other considerations (such as different security clearance) also suggest that this is not the same user role. Consequently, a separate role is defined although at implementation time both roles may use the same dialogue.

User Catalogue					SSADM Version 4		
Project/System EXAMPLE		**Author** SRS	**Date** 12/2	**Version**	**Status**	**Page** of	1 1

Job title	Job activities description
Sales Manager	Plan sales and marketing strategy
	Monitor advertising budgets and effectiveness
	Assist in planning course schedule
	Assist in design of sales material
	Manage and control sales clerks
Sales Clerk	Recording course bookings
	Respond to all enquiries
	Tele-sales established customers
Administrative Assistant	Recording course bookings
	Book hotels and conference facilities
	Collate and analyse post-course questionnaires
Training Manager	Plan course schedule
	Schedule lectures on a course

SSADMv4 45

Figure 5.7   User Catalogue for Course Administration System

In most instances there will be a one to one relationship between user role and job title. However, in small companies there may be many job holders performing the same function because of the necessary blurring of responsibilities.

User roles are documented on a standard form (Figure 5.8).

5.3.3   User Role/Function Matrix

The user perspective and the function definition are brought together on a matrix which illustrates which users can undertake which functions (see Figure 5.9(a)). Each identified cross-reference point (marked with an X) is a potential dialogue. Each row shows all the functions appropriate to a particular user. Each column identifies the user roles that may undertake that function. When more than one user role can undertake a function then the roles must be re-examined to ensure that they are distinct.

Each 'X' intersection will not necessarily become a separate dialogue. A common dialogue may be developed for different user roles where the same information presented in the same way is required for a certain function. In Figure 5.9(a) it is likely that the same dialogue will be constructed for registering booking details by the two user roles Booking Officer and Sales Manager.

Certain dialogues are critical to the development and acceptability of the system. There are particular features and issues that contribute to a dialogue being critical. For example:

- It is central to the user's business actions. In most systems there are key interfaces which help determine a user's perception of the value of the system. These are usually associated with important, perhaps even vital, business activity.

- The dialogue is frequently used. Thus the ease of the interface will again significantly affect how users view the system.

- Many user roles share the same dialogue. Hence a comprehensive dialogue has to be defined which must either be sympathetic to different user skills or adaptive to their capabilities.

- The dialogue is complex in navigation paths or error handling.

- The dialogue represents a significant change in the method of working.

Critical dialogues are circled (or may be changed to a C – see Figure 5.9(b)) and these will be used to guide technical implementation and prototype development.

This matrix can also be used as a foundation for subsequent menu development. There are many ways of grouping dialogues into menu structures and the advantages and disadvantages of each are discussed elsewhere (see References). However, the User role/Function matrix can provide a useful starting point for menu prototyping, particularly if the dialogues are to be clustered by function.

User Roles						SSADM Version 4
Project/System EXAMPLE	**Author** SRS	**Date** 14/2	**Version**	**Status**	**Page** of	l l

User role	Job title	Activities
Sales Manager	Sales Manager	Produce course schedule Monitor course bookings Registration of course bookings (5pm - 6pm only)
Sales Staff	Sales Clerks	Respond to enquiries Enquire on course status
Booking Officer	Sales Clerks Administrative Assistant	Registration course bookings
Administration	Administrative Assistant	Book hotel and conference facilities Evaluate course success
Training Manager	Training Manager	Produce course schedule Schedule lectures Enquire on course status

SSADMv4 47

Figure 5.8   User Roles for Course Administration System

User Role/Function Matrix							SSADM Version 4

Project/System	Author	Date	Version	Status	Page
EXAMPLE	SRS	15/2			of

Functions → / User roles ↓	Produce course schedule	Book hotels/facilities	Schedule lectures	Respond to enquiries	Registration of course booking	Evaluate course success	Enquire on course status	Monitor course bookings									
Training Manager	X		X				X										
Sales Manager	X				X			X									
Booking Officer					X												
Sales Staff			X				X										
Administration		X				X											

SSADMv4 46

Figure 5.9a   User Role/Function Matrix for Course Administration System

User Role/Function Matrix   SSADM Version 4

Project/System	Author	Date	Version	Status	Page 1
EXAMPLE	SRS	12/4			of 1

Functions / User roles	Produce course schedule	Book hotels/facilities	Schedule lectures	Respond to enquiries	Registration of course booking	Evaluate course success	Enquire on course status	Monitor course bookings
Training Manager	C		X				X	
Sales Manager	C				C			X
Booking Officer					C			
Sales Staff				X			X	
Administration		X				X		

SSADMv4 46

Figure 5.9b   User Role/Function Matrix showing critical dialogues

5.4   DIALOGUE DESIGN

To be effective dialogues must be:

– natural;

– consistent;

– supportive;

– non-redundant;

– flexible.

(*Coats and Vlaeminke*,1988)

Style Guides help to achieve these objectives by defining standards of good practice. These will provide guidance on layout, colour, content and error correction. Standard keys will be established for common requirements (undo, escape, help) and facilities (location of pull-down menus, selection methods, error notification and wording). In defining these interface details, style guides permit the development of consistency both within and between systems. There is no standard SSADM documentation for these guides.

5.4.1   Dialogue Structures

Dialogue Structures are based on the I/O structures. The information on the I/O Structure Descriptions are copied over onto the Dialogue Element Description forms (see Figure 5.10).

5.4.2   Logical Groupings of Dialogue Elements (LGDEs)

LGDEs define the way the user interacts with the system. These are produced from the Dialogue Structures with an LGDE being developed for each dialogue. The LGDEs effectively define steps in the dialogue and often consist of sub-functions with an input and its resultant output. Each LGDE is usually associated with a logical screen and hence a prototype (see section 5.5) can be a useful way of defining the LGDEs in association with the user. The LGDEs are superimposed on the I/O Structures as shown in Figure 5.11.

5.4.3   Dialogue Control Table

Possible paths through LGDEs are documented in a Dialogue Control Table. A default and up to three alternative pathways can be defined (see Figure 5.12). This table shows the possible paths that can be taken through the dialogue with percentage values identifying those paths which are most frequently used. LGDEs are either mandatory or optional and this can be derived from the dialogue structure. If the LGDE comprises only sequence boxes then it is mandatory (M). However, if it contains a null box it must be optional (O). LGDEs with iterations are described as M/O if the iteration can be zero. This information is added to the Dialogue Element Description form (see Figure 5.10). The iteration of Courses cannot be zero, so it is specified as mandatory. However, the

Dialogue Element Descriptions				SSADM Version 4		
Project/System EXAMPLE		**Author** SRS	**Date** 12/4	**Version**	**Status**	**Page** of

Dialogue name **Booking confirmation**

User role **Course Booking Officer**	Function **Registration of course bookings**

Dialogue element	Data item	Logical grouping of dialogue elements ID	Mandatory/ optional LGDE
CUSTOMER NUMBER	CUSTOMER ✹		
CUSTOMER DESCRIPTION	CUSTOMER NAME	BOOK CONF. 1	Mandatory
	CUSTOMER ADDRESS		
COURSE CODE	COURSE CODE		
COURSE DESCRIPTION	COURSE NAME		
	COURSE DURATION		
	COURSE COST		
COURSE RUNNINGS	COURSE RUN ✹	BOOK CONF. 2	Mandatory
	COURSE RUN DATE		
	LOCATION		
	ACCOM - FEE		
	PLACES BOOKED		
COURSE RUN No.	COURSE RUN ✹		
DELEGATE NAMES	DELEGATE NAMES	BOOK CONF. 3	Mandatory/Optional
BOOKING DETAILS	BOOK DATE		
	BOOK CLERK	BOOK CONF. 4	Optional
	BOOKING REF		

SSADMv4 9

Figure 5.10   Dialogue Element Descriptions

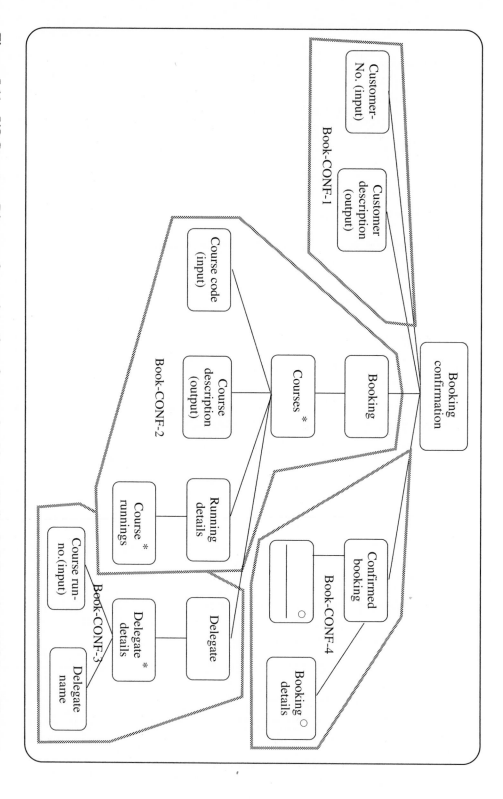

Figure 5.11   I/O Structure Diagram for registration of course bookings with Logical Grouping of Dialogue Elements

Dialogue Control Table						SSADM Version 4

Project/System EXAMPLE	Author SRS	Date 12/4	Version	Status	Page 1 of 1

Dialogue name **Booking Confirmation**

Logical grouping of dialogue elements ID	Occurrences			Default pathway	Alternative pathways		
	min	max	ave		alt 1	alt 2	alt 3
BOOK CONF. 1	1	1	1	X	. X		
BOOK CONF. 2	1	3	1	X	X		
BOOK CONF. 3	1	5	2	X			
BOOK CONF. 4	1	1	1	X			
Percentage path usage				95	5		

SSADMv4 8

Figure 5.12   Dialogue Control Table

iteration of Delegate Details may be zero if the course is full. Thus the LGDE Book-Conf-3 is declared to be Mandatory/Optional.

5.5   PROTOTYPES

A prototype can be defined as the 'first thing or being of its kind'. In information systems terms it is often a first pass design of part of the system which will later be extended and enhanced in the light of user comments and experiments.

Prototypes can be used at various points in systems development.

For example:

- to help users explore and define the problem area under consideration and to examine the feasibility of alternate designs.

- for agreeing standard interface definitions. Dialogue standards must be established for common activities such as undoing user selections, escaping from the current activity, exiting menus and screens and entering data from pull-down lists. These will be documented in a Style Guide.

- as a method of clarifying that the system developer has correctly understood user requirements. It is usually much easier for a user to evaluate a real working system with screens, menus and outputs than a paper-based specification.

- for exploring critical interfaces. The point has already been made that there are likely to be key interfaces which will largely determine the user's view of the system. Prototyping may be used to agree details of how such dialogues will be handled.

- experimenting with complex process logic to determine whether it is technically and economically viable. This may be particularly significant in systems where the performance of this process is vital for the operational success or economic viability of the system.

In core SSADM, prototyping is primarily used to demonstrate and explore aspects of the system with the user. Hence its main objective is to establish errors in specification prior to detailed design.

5.5.1   Prototype method

Prototypes may be developed in the target language of the final proposed system. The functionality and robustness of the prototype will be increased as user requirements and preferences are agreed. Hence system development progresses through a series of iterations, with each iteration bringing development nearer to an agreed solution.

Alternatively, a prototype may be written in a language designed for rapid development and offering features such as report generators, dialogue definition tools and non-procedural code, particularly relevant to prototyping. Once this prototype is agreed it effectively becomes the specification of the system, and the software is then re-written in the agreed target language. This approach is particularly appropriate in circumstances where

Prototype Demonstration Objective Document						SSADM Version 4

Project/System EXAMPLE	Author SRS	Date 12/5	Version	Status	Page of	1 / 1

Document no. 001 — Prototype pathway no. 001

Function name REGISTRATION OF COURSE BOOKING — User role BOOKING OFFICER

Agenda

1 : Demonstration and discussion of main menu and booking menu

2 : Examine how course running details will be presented

3 : Discuss how delegates will be assigned to course runnings

Component no.	Component queries

© National Computing Centre Limited, 1991

SSADMv4 37

Figure 5.13   Prototype Demonstration Objectives Document

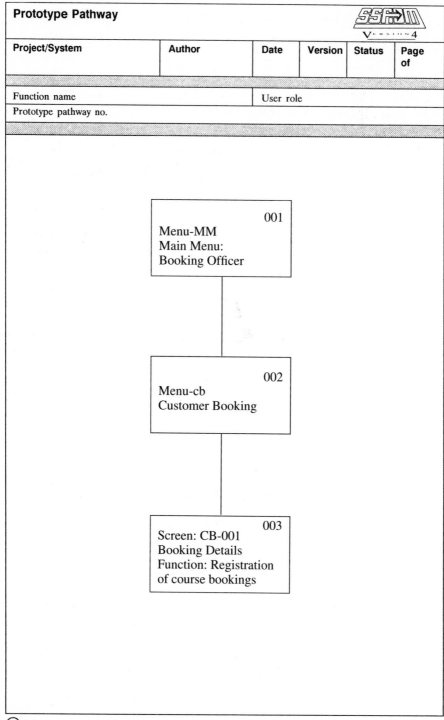

SSADMv4 38

Figure 5.14   Prototype Pathway

the installation standard is a Third Generation Language (3GL) such as COBOL or BASIC. These conventional high level languages are not particularly suitable for prototype development.

Prototyping demands that the system is written in powerful and flexible software that permits programs to be created very quickly as well as allowing easy coding and integration of extensions and amendments. Consequently, prototyping is often associated with Fourth Generation Languages (4GLs).

Perceived advantages and disadvantages of prototyping have been discussed elsewhere (see References). However, the documentation within SSADM attempts to overcome two of the problems.

1 The purpose of the prototype is not clearly understood. Alavi (1984) found that prototypes could be oversold and this led to unmet user expectations and disappointment. Hence, SSADM documentation explicitly defines the scope and objectives of the prototype so that this can be understood by both developer and user. In SSADM the objective of the prototype is documented (see Figure 5.13) and its scope defined in a pathway (see Figure 5.14).

2 The results of the prototype use are not clearly documented or agreed. Several researchers have found evidence of lack of management control in such evolutionary development and this has led to poor quality assurance and documentation. The result of the prototype demonstration is recorded in a Prototype Result Log (see Figure 5.20).

This latter point raises the worry that 'prototyping' may become a euphemism for 'throwing it together and we'll knock it up from there'. Controlled prototyping is necessary if it is to be a useful part of the development exercise.

In the following example a prototype screen has been developed to show how delegates may be allocated to a course running. This has been established as a critical dialogue and the progressive definition of the screen (Figures 5.15. to 5.19) permitted discussion of requirements and how those requirements would be handled. The result of the prototype discussion is documented in Figure 5.20.

5.6   RELATIONSHIP WITH OTHER MODELS

5.6.1   Data Flow Model

The Required System DFM provides the processes and their supporting Elementary Process Descriptions which are the starting point for establishing user initiated functions. These functions may cover part of a process, a whole process or a combination of processes (see section 5.2). The DFM will also contain the I/O Descriptions which will be extended into I/O Structures.

COURSE BOOKINGS SYSTEM – BOOKING DETAILS

Customer Code: Customer Name:

Course Code: Course Title:

Course Runnings

Run-No.   Location   Start Date   Accom. cost   Bookings
                                                           To Date.

Bookings
Run-No.   Delegate Name

Enter Customer Abbreviation, F10 For Pop-Up List

Figure 5.15   Logical Screen 1: Opening screen for Registration of Course Bookings: Prototype demonstration 001

COURSE BOOKINGS SYSTEM – BOOKING DETAILS

Customer Code: AS   Customer Name: THE ASSIST PARTNERSH
                                          WELLINGBOROUGH
Course Code: Course Title:

Course Runnings

Run-No.   Location   Start Date   Accom. cost   Bookings
                                                           To Date.

Bookings
Run-No.   Delegate Name

Enter Course Code, F10 For Pop-Up List

Figure 5.16   Logical Screen 2: Entry of customer code and retrieval of customer name (Book-CONF-1)

```
┌─────────────────────────────────────────────────────────────────────┐
│            COURSE BOOKINGS SYSTEM – BOOKING DETAILS                  │
│                                                                     │
│  Customer Code: AS        Customer Name:  THE ASSIST PARTNERSH      │
│                                           WELLINGBOROUGH            │
│  Course Code:             Course Title:                            │
│                                                                     │
│  Course Runnings                                                    │
│                           ┌──────────────────────┐       Bookings   │
│  Run-No.        Location  │ 01  BASIC SYSTEMS ANAL│  cost  To Date.  │
│                           │ 02  CLIPPER 5 - CONVER│                 │
│                           │ 03  LOTUS 123 - INTROD│                 │
│                           │ 04  SSADM - ADVANCED C │                 │
│                           │ 05  CLIPPER - ADVANCED │                 │
│                           └──────────────────────┘                 │
│  Bookings                                                           │
│  Run-No.     Delegate Name                                          │
│                                                                     │
│                                                                     │
│             Enter Course Code, F10 For Pop-Up List                 │
└─────────────────────────────────────────────────────────────────────┘
```

Figure 5.17   Logical Screen 3: Example of pop-up list for entering course code. Location and contents of the pop-up list will be documented in the style guide.

```
┌─────────────────────────────────────────────────────────────────────┐
│            COURSE BOOKINGS SYSTEM – BOOKING DETAILS                  │
│                                                                     │
│  Customer Code: AS        Customer Name:  THE ASSIST PARTNERSH      │
│                                           WELLINGBOROUGH            │
│                                                                     │
│  Course Code: 01          Course Title: BASIC SYSTEMS ANALYSIS     │
│                               3 DAY      525.00                     │
│  Course Runnings                                                    │
│                                                          Bookings    │
│  Run-No.    Location           Sart Date   Accom. cost  To Date.    │
│    1        WELLINGBOROUGH     10/MAY       80/NIGHT        6        │
│    2        MANCHESTER         15/JUNE      90/NIGHT        7        │
│    3        FRANKFURT          22/JULY      75/NIGHT        3        │
│                                                                     │
│  Bookings                                                           │
│  Run-No.     Delegate Name                                          │
│                                                                     │
│             Enter Delegate Details                                 │
└─────────────────────────────────────────────────────────────────────┘
```

Figure 5.18   Logical Screen 4: Entry of course code and retrieval of course details and runnings (book-CONF-2)

```
┌──────────────────────────────────────────────────────────────────┐
│            COURSE BOOKINGS SYSTEM – BOOKING DETAILS                │
│                                                                    │
│   Customer Code: AS        Customer Name: THE ASSIST PARTNERSH     │
│                                           WELLINGBOROUGH           │
│                                                                    │
│   Course Code: 01          Course Title: BASIC SYSTEMS ANALYSIS    │
│                                          3 DAY      525.00         │
│   Course Runnings                                                  │
│                                                           Bookings │
│   Run-No.    Location            Sart Date   Accom. cost  To Date. │
│      1       WELLINGBOROUGH      10/MAY       80/NIGHT        6     │
│      2       MANCHESTER          15/JUNE      90/NIGHT        7     │
│      3       FRANKFURT           22/JULY      75/NIGHT        3     │
│                                                                    │
│   Bookings                                                         │
│   Run-No.         Delegate Name                                    │
│      1            Tony Adams                                       │
│      1            Terry Miles                                      │
│      2            Lisa Marriot                                     │
│                                                                    │
│              Enter Delegate Details                                │
└──────────────────────────────────────────────────────────────────┘
```

Figure 5.19   Logical Screen 5: Entry of delegate details (book-CONF-3).

5.6.2   Entity-Event Modelling

Events will be defined for state and temporal activities as well as for those initiated by the user. All these events will be modelled in Entity Life Histories and Effect Correspondence Diagrams and must be recorded within function definitions. The relationship of events to functions was also considered in section 5.2. The construction of the Entity Life Histories will significantly tighten up the initial Function Descriptions developed from the DFM.

5.6.3   Enquiry Access Paths

Enquiry requirements are recorded in the Requirements Catalogue during the construction of the Data Flow Model (DFM) and Logical Data Model (LDM). However, no I/O descriptions are constructed for these enquiries and hence it is only during function definition that they are defined in more detail. The I/O Structures which describe each enquiry are used as a basis for the Enquiry Access Path, which in turn is used to validate the Logical Data Model (see Chapter 8).

Function definition and dialogue modelling will provide important inputs into Logical Database Process Design (Chapter 9) and subsequent Physical Process Specification (Chapter 11).

Prototype Result Log						*SSADM* Version 4	
Project/System **EXAMPLE**		Author **SRS**	Date **13/5**	Version	Status	Page **1** of **1**	

Prototype result log no. **001**   Prototype pathway no. **001**

Function name **REGISTRATION OF COURSE BOOKING**   User role **BOOKING OFFICER**

Component no.	Result no.	Result description	Change grade
001	1	Rewording of "client list" to "customer list"	
002	1	Slight mispelling in demo menu	
003	1	Extend customer name field to show all of name	
	2	Currency of accomodation costs needs showing. Frankfurt figure should relate to DM	
	3	System should allow up to 8 delegates / booking	
	4	Where is the booking number coming from? Should it be displayed?	

© National Computing Centre Limited, 1991   SSADMv4 39

Figure 5.20   Prototype Log

5.7   SUMMARY

This chapter has examined three inter-related aspects of SSADM development; Function Definition, Dialogue Design and Prototyping. These are linked by their focus on the user's perspective of the system and the acknowledgement that the way activities are grouped, presented and structured contributes significantly to the user's perception of the value of the system. Examples of representative documentation have been provided during the course of the chapter.

SSADM Version 3 did not specifically address functional definition and prototyping but used Logical Dialogue Outlines (LDOs) to represent dialogue design. These essentially flowcharted the dialogue using a limited set of symbols (see Figure 5.21). LDOs were not particularly popular, perhaps partly due to their restricted definition within SSADM and also their perceived irrelevance to 'screen-at-a-time processing'. However, one of the authors of this book has used an extended notation set in a number of projects and they have been welcomed by developers and users alike (Skidmore,1990). They do address aspects of dialogue routing and control, an issue that requires closer examination in the SSADM Version 4 context.

5.7.1   Dialogue Controls

The SSADM Reference Manual suggests that dialogue error handling should not be extensively addressed in dialogue design. Function Definitions will record general comments on error procedure and "any useful information is noted". However,

> "The entity-event modelling and logical process design techniques . . . tackle and document the error handling formally."

Practitioners familiar with Logical Dialogue Outlines (LDOs) from SSADM Version 3 and State Transition Diagrams (see Further Reading) will be aware of the advantages of defining dialogue control at an early stage. It is particularly useful to do this in highly interactive systems developed using software that features WIMP and GUI interfaces.

The Dialogue Structures can be easily extended using the operation nodes given in the SSADM Structure Diagram (see Chapter 2) and an example is given in Figure 5.22. It must be stressed that this enhancement is not defined within the SSADM reference manual but it seems to us to be within the spirit of the use of the Structure Diagram.

Finally, the iterative development of Function Definitions needs re-statement. The SSADM Reference Manual describes it as a 'super-technique'which involves an iterative group of activities. It is not a technique like logical data modelling or data flow diagramming but a procedure for bringing together existing products and providing a user-centred reference point.

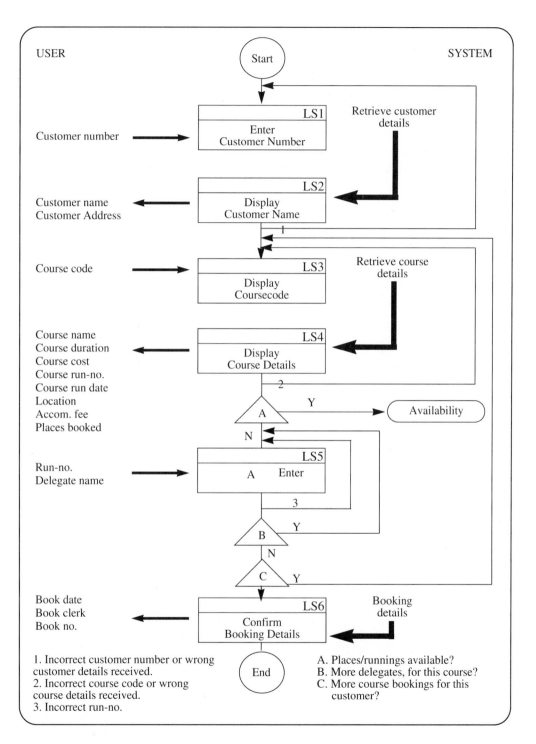

Figure 5.21 Logical Dialogue Outline for Registration of Course booking.

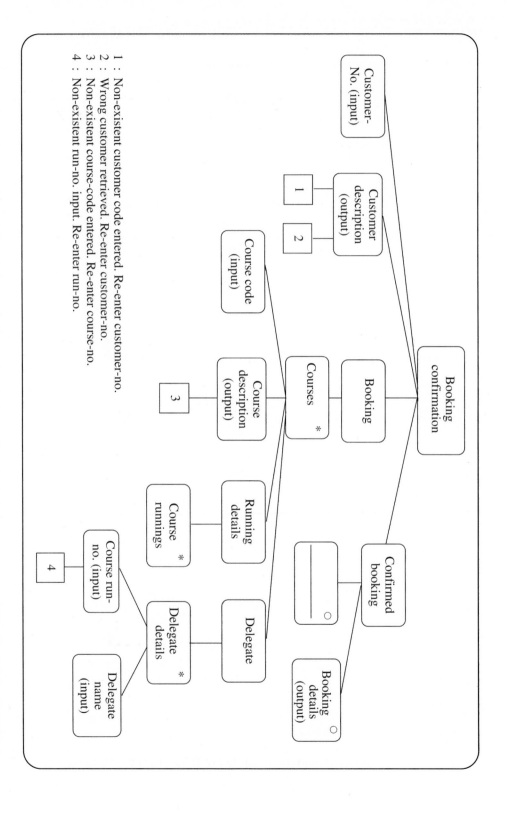

1 : Non-existent customer code entered. Re-enter customer-no.
2 : Wrong customer retrieved. Re-enter customer-no.
3 : Non-existent course-code entered. Re-enter course-no.
4 : Non-existent run-no. input. Re-enter run-no.

Figure 5.22   I/O Structure Diagram for Registration of course bookings with operation notes for dialogue control

6   Relational Data Analysis

6.1   INTRODUCTION

Relational Data Analysis (RDA) is a technique for deriving data structures which are unambiguous, free from redundant duplication, flexible and logically easy to maintain. The technique involves applying a series of 'normalisation rules' to groups of data items. The resultant data groupings are known as normalised relations.

A collection of normalised relations constitutes a relational model and can be represented diagrammatically in the form of a Logical Data Model.

Relational Data Analysis may be used:

- in the analysis of the current environment to identify data contents and relationships between data items;

- as part of the process of creating the Required System LDM:

  - informally as the LDM is developed;

  - formally by converting LDM to relations to check it is normalised.

- to verify the LDM against the Function Definitions by selective analysis of I/O Structures and I/O Structure Descriptions.

Type Code	Description	Catalogue No.	Size
SSU	Single Sided Uprights	9011	43cm (16¾in)
		9012	71cm (28in)
		9014	122cm (48in)
		9018	198cm (78in)
DSU	Double Sided Uprights	9608	7.6cm x 2.5cm x 300cm
			(3in x 1in x 118in)
		9221	2.5cm x 3.3cm x 300cm
			(1in x 1⅜in x 118in)
SB	Straight Brackets	9002	17cm (7in)
		9002B	22cm (9in)
		9003	27cm (10½in)
SLB	Slanting Brackets	9022	18cm (7¼in)
		9022B	23.5cm (9½in)
		9023	28cm (11¼in)

Figure 6.1   Tabular representation of Product Data

6.2   MODELLING CONCEPTS

The concepts underlying Relational Data Analysis rely on the tabular representation of data. Everyday examples of this format are bus or railway timetables. Another example is shown in Figure 6.1 where product details are listed in a catalogue.

6.2.1   Relation

A relation is a two-dimensional table (Figure 6.2) comprising of a number of columns and a number of rows of data. Each column represents an attribute (data item) of the relation.

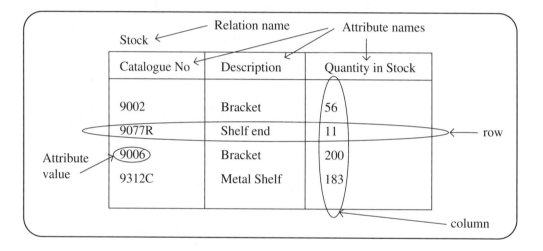

Figure 6.2   Stock Relation

To be a relation the table must have the following properties:

– no two rows are the same. This prohibits two rows from having the same attribute values throughout. Thus a row can always be uniquely identified by quoting an appropriate combination of attribute values.

– the order of the rows is not significant. If row order is significant it indicates that there is some meaning in the sequence and that there is data missing from the relation. For example, a bus timetable lists the names of the scheduled stops. For the timetable to be represented as a relation a bus stop sequence number must be included as an additional attribute to indicate the sequence of the bus stops.

– the order of the columns is not significant. Columns can be interchanged without affecting the information content of the relation. This can be enforced by ensuring that each column has a unique name. If two columns in a table have the same name and the attribute values are drawn from the same domain, each must have its 'role' included in its name to ensure uniqueness. A relation listing married couples where each partner is identified by the attribute National-insurance-number would be ambiguous unless the columns were labelled Husband-national-insurance-number and Wife-national-insurance-number.

In addition, for a relation to be Normalised:

- all attributes are atomic – each row/column intersection contains a single attribute value.

6.2.2 Domain

A domain represents a 'pool' of values from which the values for attributes are drawn. A domain can be identified for individual attributes or a group domain for a group of attributes. A domain forms a method of documenting the validation rules, formatting rules, permitted classes and ranges of values that are associated or common to the attributes in the domain. Domains may also be useful in the identification of homonyms and synonyms. For example, ascertaining whether Requisition-number and Order-number are from the same domain may help to decide whether the attribute names are synonyms or are actually two different attributes. Likewise, recognising that an Order-number may be drawn from different domains may help identify the distinction between Purchase-order-number and Sales-order-number, and hence the relations Purchase Order and Sales Order.

6.2.3 Keys

Each row of a relation is distinct, hence the value of an attribute or collection of attributes can be used to uniquely identify any row in the relation. A candidate key is an attribute, or group of attributes, which can never have duplicate values within an occurrence of a relation, and whose value is therefore always sufficient to identify a row. A candidate key is always minimal in the sense that no sub-group of attributes of the candidate key can itself be a candidate key.

The candidate key selected to be the unique identifier of a relation is known as the primary key.

A candidate key, normally the primary key, of one relation which is included in another relation in order to represent a relationship between the two relations is known as a foreign key.

A foreign key is indicated by an asterisk (*) in a normalised relation.

A candidate key which consists of only one attribute is known as a simple key. A compound key consists of two or more attributes.

A candidate key which includes, but is not entirely made up of foreign keys, is known as a composite or hierarchical key.

The distinction between these types of key is introduced in context within this chapter.

6.2.4 Normalisation

Normalisation is the process of transforming the data through a series of normal forms to produce a set of simple, unambiguous relations which are free of redundant data and undesirable features. Redundant data occurs when data is unnecessarily duplicated in the system. The undesirable features caused by redundant data include additional processing of duplicated data, potential inconsistency of data and update anomalies (see section 6.10).

The concept of normalisation of data was developed by Dr E F Codd in a series of seminal papers, the first in 1970 (see References). The technique has a strong mathematical background based on the theory of relations. Codd further developed his ideas refining the concept of the relational model, leading to the emergence of relational database theory.

6.2.5 Repeating Groups

In many tabular examples of data there are instances of 'repeating groups'. For example, Figure 6.1 shows a product list in which each product type is available in a number of sizes. So for each product type there is a repeating group of catalogue number and size information.

A repeating group can be considered as one relation nested inside another. In this example, relations with the size details are nested inside the relation giving the product types. The identification of repeating groups is important to the process of RDA. Repeating groups can be recognised on the I/O Structure Diagrams as iterations (see Chapter 5).

Product

Type-code

Description

    Catalogue Number

    Metric-size

    Imperial-size

Figure 6.3 Representation of Product Data as a list

6.2.6 Notation for Relations

A relation can be represented in tabular form, as in Figure 6.2. Such a table represents a snapshot of the data at one instance of time. For the purpose of Relational Data Analysis, a relation is better represented by the relation name followed by a list of the attribute names, as in Figure 6.3, where:

  – the Primary key of the relation is shown by underlining each of its component attributes;

- the attributes of any repeating groups (nested relations) within the relation are indented;
- a candidate key for each repeating group is underlined.

6.2.7   Functional Dependency

The formal definition of functional dependency is:

> An attribute Y of a relation R is functionally dependent on another attribute X of R if and only if each value of X is associated with only one value of Y.

For example, Department-name is functionally dependent upon Employee-code if and only if each value of Employee-code is associated with only one value of Department-name. However, an Employee-code is not functionally dependent on Department-name because there will normally be more than one employee working in any one department. A simple way of showing functional dependencies is to draw a functional dependency diagram as in Figure 6.4.

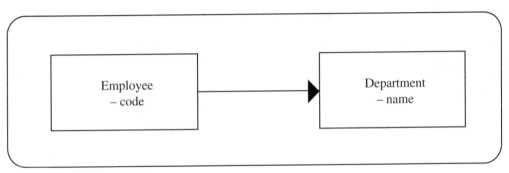

Figure 6.4   Functional Dependency Diagram

Establishing functional dependencies is part of the process of understanding what data means. The fact that Department-name is functionally dependent on Employee-code comes from the recognition that each employee works in precisely one department. This functional dependency reflects a business rule of the enterprise and must be validated by the users of the system. Many system problems are caused by the developer of the computer system assuming or imposing functional dependencies which do not reflect actual business operations and requirements.

So, for example, if it is necessary to keep a record of all the departments that an employee has worked in since joining the company, then Department-name would be functionally dependent on Employee-code and Transfer-date combined. Figure 6.5 shows that Department-name is fully functionally dependent on both the attributes Employee-code and Transfer-date and not on any subset of the group.

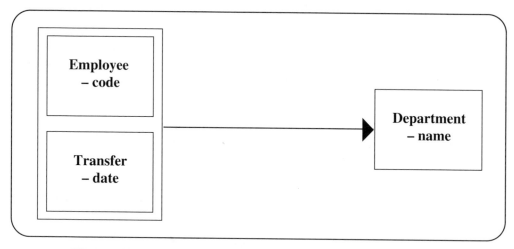

Figure 6.5　Full Function Dependency on a group of attributes

A partial, or part key, functional dependency is shown in Figure 6.6, where Employee-name is dependent on only a subset of a group of attributes, in this case Employee-code.

It is not necessary to represent all functional dependencies diagrammatically. However it may be desirable to select instances where there appears to be complex inter-relationships between the attributes. They can also be an economic and unambiguous method of documenting business system rules.

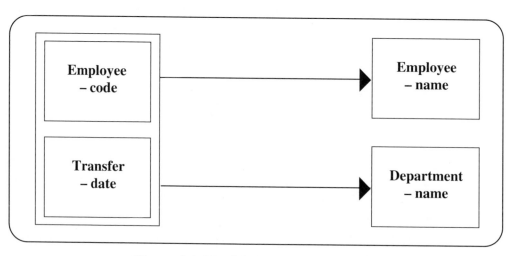

Figure 6.6　Partial Functional Dependency

6.2.8   Determinant

A determinant is defined to be any attribute, or group of attributes, on which some other attribute is fully dependent. In Figure 6.6, Employee-code is a determinant of Employee-name while Employee-code and Transfer-date together is the determinant of Department-name. The concept of determinant is useful in recognising candidate keys of a relation.

6.3   CONSTRUCTING THE MODEL

Relational data analysis is concerned with the normalisation of the logical data. By examining the dependencies between attributes, it is possible to determine whether a group of attributes represents a single relation or a number of relations.

The steps involved in producing Third Normal Form (3NF) relations are:

– represent unnormalised data;

– convert to First Normal Form (1NF);

– understand the dependencies;

– convert to Second Normal Form (2NF);

– convert to Third Normal Form (3NF);

– rationalise results.

To illustrate the normalisation process we will consider a report showing installation and servicing of machines. The report is structured to show the account details, order details for machine installation (including the installing engineers and machine details), each subsequent service of the machines and the replacement parts supplied, and finally the payments received from the account. The structure of the report is shown in Figure 6.7.

```
                    Account details
                 Installation order details
                  Installation engineers
                     Machine details
                 Service details
                    Machines serviced
                       Parts supplied details
                 Payments received
```

Figure 6.7   Structure of installation and service report

6.3.1   Represent Unnormalised Data as a List of Attributes

This process requires an initial understanding of the data and the structure of the data in order to:

- list the attributes;
- identify the repeating groups;
- underline the key attributes:
  - primary key for the un-normalised data;
  - a primary key for each repeating group.

Listing the attributes to be included in the analysis requires an understanding of the source of the data and its use within the application. Physical data such as pointers, end of field/record markers, length of field/record data, error status, overflow indicators, report and form data such as page numbers, heading and report identification items including dates, is not included in the analysis process.

Derived data must be included in the analysis but derivable attributes may be excluded if this is agreed as a local standard. The distinction between derived and derivable data is important. An invoice total may be derivable from the individual line totals and is printed on the invoice only for the convenience of the user. The line total is derived, at its simplest, from the unit cost of the item at the time of the order and the quantity invoiced for. Only if a complete history of unit costs for each item is retained in the data can the line total be derivable at any time. Identifying which data is derived and which will continue to be derivable is important in understanding the business aspects of the data.

Including all derivable data in the relational data analysis process can lead to a proliferation of attributes and some confusion when functional dependencies are considered. However, if the rules for the retention of data are not clear or not yet agreed then it may be prudent to include them in the analysis process.

Whether included or not, all derivable attributes must be described on an Attribute/Data Item Description form specifying their derivation from other attribute values.

Figure 6.8 shows the unnormalised data listed for the Account example. Each level of repeating group is indented. Alternatively, the data can be listed on an RDA Working Paper (Figure 6.9) where the level number indicates the depth of nesting for the repeating groups. All derivable data has been excluded from this example. Each machine is identified by a compound attribute of its machine type and a serial number associated with that machine type.

6.3.2   Convert to First Normal Form (1NF)

The process of converting to 1NF consists of partitioning the relation(s) by:

- removing repeating groups;
- propagating the higher level primary keys;
- assigning meaningful names to each new relation.

Each repeating group is removed from the unnormalised list of data to form a relation on its own. To retain a link the primary key of the residual, or higher order relation, is

Account
Account-number
Company-name
Company-address
Telephone-number
Contact-name

    Order-number
    Order-date
    Installation-date
    Invoice-number
    Invoice-amount

        Installation-engineer-number
        Engineer-name

        Machine-type
        Machine-serial-number
        Machine-characteristics
        Machine-description

    Service-invoice-number
    Service-date
    Service-engineer-number
    Engineer-name
    Invoice-amount

        Machine-type
        Machine-serial-number
        Service comments

            Part-code
            Part-description
            Quantity

    Transaction-reference
    Payment-received-amount
    Payment-date

Figure 6.8   Unnormalised data list, indented to show repeating groups

RDA Working Paper							*SSADM* Version 4
Project/System			Author	Date	Version	Status	Page of

Source name   **Account**

UNF		1NF	2NF	3NF	Result	
Attributes	Level				Relation	Attributes
Account·number	1					
Company·name	1					
Company·address	1					
Telephone·number	1					
Contact·name	1					
Order number	2					
Order date	2					
Installation date	2					
Invoice number	2					
Invoice amount	2					
Installation engineer number	3					
Engineer name	3					
Machine type	3					
Machine serial number	3					
Machine—characteristic	3					
Machine description	3					
Service Invoice number	2					
Service date	2					
Service engineer number	2					
Engineer name	2					
Invoice amount	2					
Machine Type	3					
Machine Serial number	3					
Service comments	3					
Part code	4					
Part description	4					
Quantity	4					
Transaction reference	2					
Payment received	2					
Payment date	2					

SSADMv4 40

Figure 6.9   Listing the unnormalised data

included in each of the new relations to form an hierarchic (composite) key. This is shown in Figure 6.10, where the three level 2 repeating groups have become three new relations, Installation Order, Service and Transaction. Each includes the Account primary key as part of their primary key.

Installation Order
Account-number
Order-number
Order-date
Installation-date
Invoice-number
Invoice-amount

 Installation-engineer-number
 Engineer-name

 Machine-type
 Machine-serial-number
 Machine-characteristics
 Machine-description

Account
Account-number
Company-name
Company-address
Telephone-number
Contact-name

Service
Account-number
Service-invoice-number
Service-date
Service-engineer-number
Engineer-name
Invoice-amount

 Machine-type
 Machine-serial-number
 Service comments

 Part-code
 Part-description
 Quantity

Transaction
Account-number
Transaction-reference
Payment-received-amount
Payment-date

Figure 6.10   First partitioning of the unnormalised data

Two of the new relations, Installation Order and Service, still contain repeating groups. The process needs to be repeated once more for Installation Order and twice more for Service to give the set of 1NF relations in Figure 6.11.

The resulting 1NF relations have 'primary keys' which are not necessarily minimal, in that not all the attributes need to be included in the key in order to uniquely identify a row. This is caused by the mechanistic process used to derive the 1NF relations. These anomalies will be resolved as the functional dependencies are considered.

Account
Account-number
Company-name
Company-address
Telephone-number
Contact-name

Installation Order
Account-number
Order-number
Order-date
Installation-date
Invoice-number
Invoice-amount

Installation Engineer
Account-number
Order-number
Installation-engineer-number
Engineer-name

Installed Machine
Account-number
Order-number
Machine-type
Machine-serial-number
Machine-characteristics
Machine-description

Service Call
Account-number
Service-invoice-number
Service-date
Service-engineer-number
Engineer-name
Invoice-amount

Service
Account-number
Service-invoice-number
Machine-type
Machine-serial-number
Service-comments

Part Supplied
Account-number
Service-invoice-number
Machine-type
Machine-serial-number
Part-code
Part-description
Quantity

Transaction
Account-number
Transaction-reference
Payment-received-amount
Payment date

Figure 6.11   Relations in 1NF

6.3.3   Understand the Dependencies

For each of the 1NF relations it is necessary to understand the functional dependencies to decide whether the relations are already in 2NF or need to be further partitioned. When deriving functional dependencies the users should be involved and assumptions should not be made. Figure 6.12 shows the functional dependency diagrams for each of the 1NF relations given in Figure 6.11.

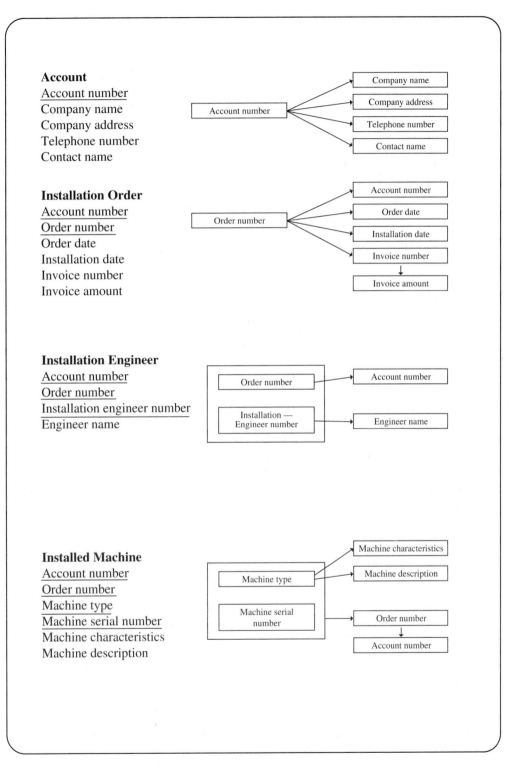

Account
Account number
Company name
Company address
Telephone number
Contact name

Installation Order
Account number
Order number
Order date
Installation date
Invoice number
Invoice amount

Installation Engineer
Account number
Order number
Installation engineer number
Engineer name

Installed Machine
Account number
Order number
Machine type
Machine serial number
Machine characteristics
Machine description

Figure 6.12 Functional dependency diagram for the INF relations

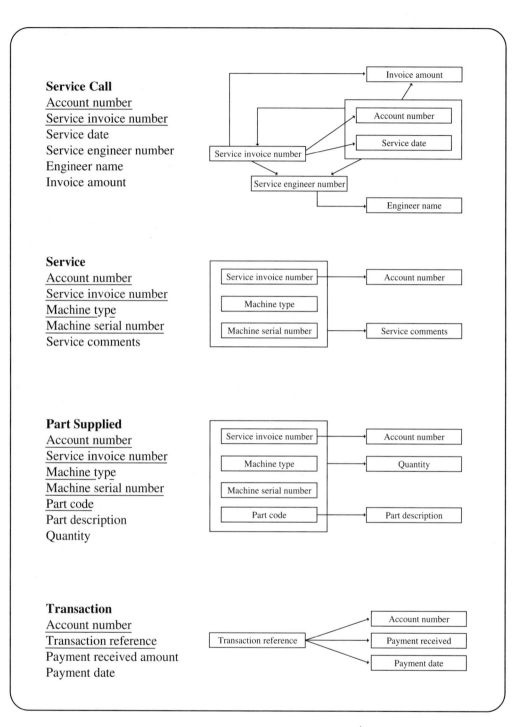

Figure 6.12 (cont)   Functional dependency diagrams for the 1NF relations

1NF	2NF
Account	No change
Account-number	
Company-name	
Company-address	
Telephone-number	
Contact-name	
Installation Order	**Installation Order-2**
Account-number	Account-number
Order-number	Order-number
Order-date	Order-date
Installation-date	Installation-date
Invoice-number	Invoice-number
Invoice-amount	Invoice-amount
Installation Engineer	**Account Order-2**
Account-number	Order-number
Order-number	Account-number
Installation-engineer-number	
Engineer-name	**Engineer-2**
	Installation-engineer-number
	Engineer-name
	Installation Engineer-2
	Order-number
	Installation-engineer-number
Installed Machine	**Installed Machine-2**
Account-number	Account-number
Order-number	Order-number
Machine-type	Machine-type
Machine-serial-number	Machine-serial-number
Machine-characteristics	
Machine-description	**Machine Type-2**
	Machine-type
	Machine-characteristics
	Machine-description
Service Call	**Service Call-2**
Account-number	Account-number
Service-date	Service-invoice-number
Service-invoice-number	Service-date
Service-engineer-number	Service-engineer-number
Engineer-name	Engineer-name
Invoice-amount	Invoice-amount

Figure 6.13 Identifying the minimal primary key and removing part key dependencies from 1NF relations to give 2NF relations

```
        1NF                              2NF
        Service                          Service Invoice-2
        Account-number                   Service-invoice-number
        Service-invoice-number           Account-number
        Machine-type
        Machine-serial-number
        Service-comments


                                         Machine Service-2
                                         Service-invoice-number
                                         Machine-type
                                         Machine-serial-number
                                         Service-comments


        Part Supplied                    Part Supplied-2
        Account-number                   Service-invoice-number
        Service-invoice-number           Machine-type
        Machine-type                     Machine-serial-number
        Machine-serial-number            Part-code
        Part-code                        Quantity
        Part-description
        Quantity
                                         Part-2
                                         Part-code
                                         Part-description

                                         Service Invoice-2a
                                         Service-invoice-number
                                         Account-number


        Transaction                      Transaction-2
        Account-number                   Account-number
        Transaction-reference            Transaction-reference
        Payment-received-amount          Payment-received-amount
        Payment date                     Payment date
```

Figure 6.13 (cont)   Identifying the minimal primary key and removing part key dependencies from 1NF relations to give 2NF relations

6.3.4   Convert to Second Normal Form (2NF)

Deriving 2NF relations from the 1NF relations is a two step process:

- Consider the functional dependencies and identify the candidate keys for each relation. Select one to be the primary key.

- Remove any part key dependencies by partitioning the relations.

The partitioning of the 1NF relations is illustrated in Figure 6.13. The Account relation already has a minimal key on which all its attributes are functionally dependent. Consequently it is already in 2NF.

A sufficient primary key for the relation Installation Order is Order-number since it determines all the other attributes. The new 2NF relation is called Installation Order-2 and all the attributes are functionally dependent on the new primary key.

It can be seen from the dependency diagram for the relation Installation Engineer that the minimal primary key is Order-number, Installation-engineer-number. This gives two part key functional dependencies. Each Account-number is dependent on Order-number and Engineer-name is dependent on Installation-engineer-number. When these part key dependencies are removed into Account Order-2 and Engineer-2 relations respectively the Installation Engineer-2 relation remains as a key only relation.

The Service Call relation has two candidate keys, Account-number, Service-date and Service-invoice-number. In consultation with the user, it is decided that Service-invoice-number is the more appropriate primary key and this is adopted in the Service Call-2 relation.

Each 1NF relation with its functional dependency diagram is separately considered and 2NF relations derived. Thus only a few attributes and their associations are being examined at any one time. This reduces the massive task of data analysis to manageable 'chunks' that are reasonably easy for both user and analyst to comprehend.

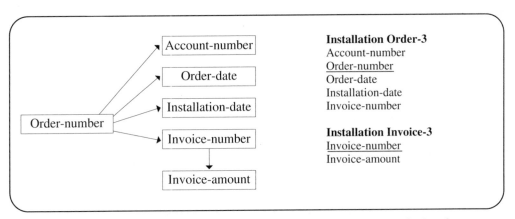

Figure 6.14   Dependency diagram for 2NF Installation Order-2 and its partitioning into 3NF relations

6.3.5   Convert to Third Normal Form (3NF)

For a relation to be in 3NF all the non-key attributes should be fully functionally dependent on the primary key of the relation. That is, the primary key of the relation determines all the non-key attributes in the relation, and non-key attributes of the relation are not determinants unless they are themselves a candidate key for the relation.

The functional dependency diagrams can be used to identify determinants which represent candidate keys of relations. For example, the functional dependency diagram for

Account
Account-number
Company-name
Company-address
Telephone-number
Contact-name

Installation Order-3
Order-number
Account-number
Order-date
Installation-date
Invoice-number

Account Order-2
Order-number
Account-number

Account Order-3
Order-number
Account-number

Engineer-2
Installation-engineer-number
Engineer-name

Service Engineer-3
Service-engineer-number
Engineer-name

Installlation Engineer-2
Order-number
Installation-engineer-number

Installed Machine-3
Machine-type
Machine-serial-number
Order-number

Machine Type-2
Machine-type
Machine characteristics
Machine-description

Service Call-3
Service-invoice-number
Account-number
Service-date
Service-engineer-number
Invoice-amount

Service Invoice-2
Service-invoice-number
Account-number

Service Invoice-2a
Service-invoice-number
Account-number

Machine Service-2
Service-invoice-number
Machine-type
Machine-serial-number
Service-comments

Installation Invoice-3
Invoice-number
Invoice-amount

Part-2
Part-code
Part-description

Part Supplied-2
Service-invoice-number
Machine-type
Machine-serial-number
Part-code
Quantity

Transaction-2
Transaction-reference
Account-number
Payment-received-amount
Payment date

**Figure 6.15   The 3NF relations arranged to group together identical
primary keys, ready for rationalisation**

Account, given in Figure 6.12, shows Account-number to be the determinant of all the attributes in the relation. Thus Account is not only in 2NF but in 3NF also.

The functional dependency diagram for Installation Order-2, Figure 6.14, shows two determinants, Order-number and Invoice-number. Only Order-number is a candidate key for the 2NF relation. To remove the non-candidate key dependency, the relation is partitioned by removing the dependent attributes and their determinant into a new relation, Installation Invoice-3. Invoice-number also remains in the original relation, now named Installation Order-3, to maintain the link.

Each 2NF relation is considered in turn and partitioned into 3NF relations if they are not already in 3NF. Consequently, Installed Machine-2 is partitioned into Installed Machine-3 and Account Order-3, and Service Call-2 is partitioned into Service Call-3 and Service Engineer-3.

During the normalisation process the primary keys have been underlined. It is also convention to place the primary key attributes first in the attribute list. The resulting set of 3NF relations is shown in Figure 6.15.

6.3.6   Rationalise Results

The process of deriving the 3NF relations has resulted in a set of relations which contain duplication. To rationalise this set of relations it is necessary to:

- consider whether to combine any resulting relations that have identical primary or candidate keys;

- discard any relations that are redundant. A relation is redundant if its attributes are contained within another relation.

For example, Figure 6.15 shows that Account Order-2 and Account Order-3 are identical and both contained in the 3NF relation Installation Order-3. Both relations can be rationalised out of the set.

Service Invoice-2 and 2a can be combined into Service Call-3 and no information would be lost. Furthermore, on checking with the users it is found that Invoice-number, the primary key of Installation Invoice-3, is drawn from the same domain as Service-invoice-number. It could also be combined with Service Call-3 and the relation more appropriately named. However, there are disadvantages to this combining of relations. Each row of the relation representing installation invoices would contain inappropriate attributes and this would be wasteful of storage space. This could be outweighed by the processing advantages of only having one relation containing invoice data. Other factors which might be considered could be the number of installations compared to the number of services. Alternatively, the three invoice relations could be merged into a single Invoice relation, removing Invoice-amount from Service Call-3 to avoid redundant data. Each of these solutions would be in 3NF. However, after discussion, the user has decided that keeping the two types of invoices in separate relations is more appropriate to the business requirements.

Account
Account-number
Company-name
Company-address
Telephone-number
Contact-name

Installation Order
Order-number
*Account-number
Order-date
Installation-date
*Invoice-number

Engineer
Engineer-number
Engineer-name

Installlation Engineer
Order-number
Installation-engineer-number

Installed Machine
Machine-type
Machine-serial-number
*Order-number

Machine Type
Machine-type
Machine characteristics
Machine-description

Service Invoice
Service-invoice-number
*Account-number
Service-date
*Service-engineer-number
Invoice-amount

Installation Invoice
Invoice-number
Invoice-amount

Part
Part-code
Part-description

Transaction
Transaction-reference
*Account-number
Payment-received-amount
Payment date

Machine Service
Service-invoice-number
Machine-type
Machine-serial-number
Service-comments

Part Supplied
Service-invoice-number
Machine-type
Machine-serial-number
Part-code
Quantity

Figure 6.16   The 3NF relations rationalised and renamed as appropriate

The 3NF relations Engineer-2 and Service Engineer-3 appear to hold the same information and could therefore be merged into one relation Engineer. Before this is approved, however, it would be necessary to check that the corresponding attribute values in each relation are drawn from the same domain and to ensure that no information is lost in this merger. It would be important to check whether all engineers could both install and service machines. If a distinction is required between installation engineers and service engineers, then either the two relations will have to be retained or an additional skill attribute added to the merged relation. In our example there is no distinction between installation engineers and service engineers and the relations are merged. Note that it is not necessary to change the instances of attribute names to correspond, only to remember that that Service-engineer-number, Installation-engineer-number and Engineer-number are all synonyms drawing values from the same domain.

Once the set of 3NF relations is established each foreign key, which is not part of a primary key, is marked with an asterisk (*).

Finally the names of the rationalised relations are reviewed to ensure that the relation name reflects the information to be held by the relation. If the relations are to be compared with the LDM, the name should be consistent with corresponding entities on the LDS. The rationalised set of 3NF relations, renamed as appropriate, are shown in Figure 6.16.

6.4   REPRESENTING 3NF RELATIONS AS A LOGICAL DATA MODEL

A Logical Data Model and a set of normalised relations are two different ways of modelling similar data.

- entities in the LDM correspond to relations;
- each occurrence of an entity represents a row in a relation;
- relationships in the LDM are a representation of the foreign key correspondence of the relations.

During the development of the LDM it is possible to validate the model by comparing it with sub-models produced from normalising I/O Structures, reports and other samples of data. To do the comparison it is usual to represent the 3NF relations in the form of a relational data sub-model.

The rules for converting a set of 3NF relations to a LDM sub-model are:

- create an entity type for each relation;
- mark the higher level elements of hierarchic keys as foreign keys;
- check that all the masters of compound key relations are present;
- make compound key relations into details;
- make relations with foreign keys into details.

Once the two models are in the same form they may be compared.

The construction of a logical data sub-model is illustrated using the set of 3NF relations developed in the previous sections.

6.4.1   Create an Entity Type for Each Relation

Each 3NF relation is represented on the diagram as an entity type and named with the relation name. It is also useful to include the primary key (underlined) and any foreign keys (marked by *) in the entity box. This will aid the application of the remaining rules.

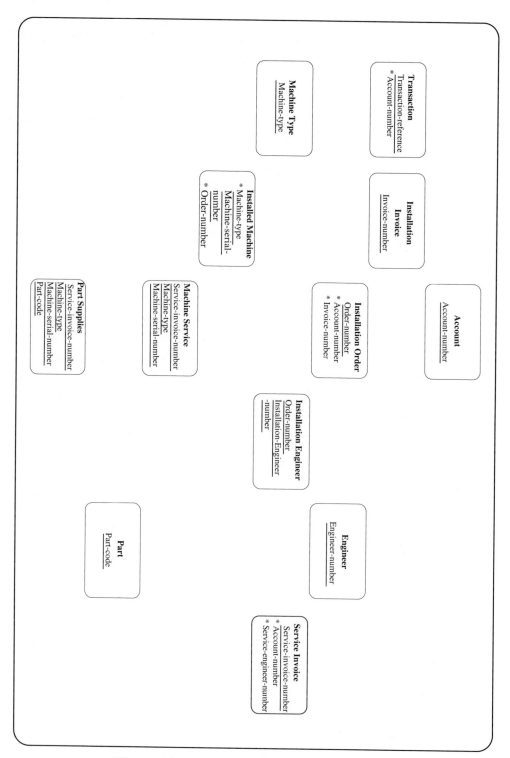

Figure 6.17   Relations represented as entities

Whenever possible, it is useful to place the entities in a similar layout to their equivalent entities on the LDM.

6.4.2   Mark the Higher Level Elements of Hierarchic Keys as Foreign Keys

If the entire primary key of a relation is a hierarchic (composite) key, then the higher level elements are marked as a foreign key. These are the attributes which were the key of the outer (or higher) relation and were copied into a nested relation when developing 1NF relations. Installed Machine has the only hierarchic key in our example, so Machine-type is marked as a foreign key.

The application of the above two rules is illustrated in Figure 6.17.

6.4.3   Check that All the Masters of Compound Key Relations are Present

Check that each attribute of every compound key occurs as the simple or hierarchic key of another relation. If an element is part of a compound key, but not a simple key of another relation then:

– create a new entity type with the attribute as its primary key;

– make this entity the master of each entity which has the attribute as part of its compound key;

– mark it as a foreign key in all other relations where it appears as a non-key attribute.

The relations Installation Engineer, Machine Service and Part Supplied have compound keys. However, each component of the key does occur as a simple or hierarchic key of another relation

6.4.4   Make Compound Key Relations into Details

Entity types with compound keys are made the details of the entity types that have an attribute or group of attributes of the compound as their total prime key. Each element must be allocated only once.

Thus, in Figure 6.18, it can be seen that:

– Installation Engineer becomes the detail of Installation Order and the detail of Engineer.

– Machine Service becomes the detail of Installed Machine and the detail of Service Invoice.

– Part Supplied becomes the detail of Machine Service and the detail of Part.

6.4.5   Make Relations with Foreign Keys into Details.

A relation with a foreign key is made the detail of the relation that has foreign key as its

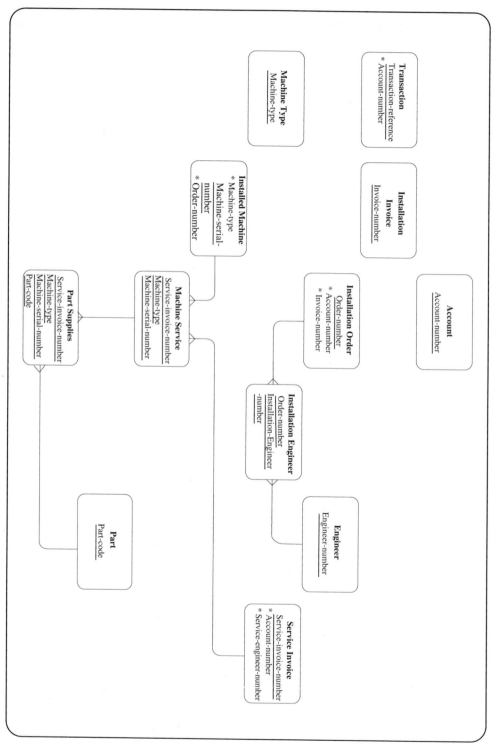

Figure 6.18   Compound key relations become details

complete prime key. Figure 6.19 shows the application of this rule.

Although the application of the above rules appears straightforward and mechanistic, this is not the case in reality. The resulting relational data sub-model needs to be checked and a number of issues considered.

6.5   ISSUES IN THE LDM REPRESENTATION OF RELATIONS

6.5.1   One to One Relationships

The process of identifying relationships by the foreign keys has led to a set of one to many (master-detail) relationships. Consideration must now be given to each relationship to see if any are, in fact, one to one (1:1) relationships in the model. In our example the user assures us that each Installation Order generates only one Installation Invoice and that the invoice will contain the details of only one installation order. The relationship can be shown as 1:1 but in such circumstances care should be taken to enquire about procedures for dealing with incorrect invoices. Such procedures could cause more than one invoice to exist for each order. In this instance we will take the user's assurance that this is not the case and modify the sub-model to show a 1:1 relationship.

6.5.2   Optional Relationships

A foreign key contained within the primary key of a relation indicates a mandatory relationship, as no part of a primary key can ever be null. Nevertheless, for foreign keys which are not part of primary keys we must determine whether they can have null values. If they can, then the relationship represented by that foreign key is optional.

In our example a number of optional relationships can be identified. An Account can exist before any Transactions have been received or Service Invoices issued, but not before an Installation Order is made. Similarly, parts may exist which have not yet been supplied on a machine service. Figure 6.20 shows the relational data sub-model of Figure 6.19 modified to show the 1:1 and optional relationships.

6.5.3   Multiple Level Hierarchies

The hierarchy of Machine Type, Installed Machine, Machine Service and Part Supplied shows a 'cascading' of attributes to form the primary keys. In applying the rule 'make compound key relations into details', questions need to be asked about the relationships between individual elements of the compound keys. In our example the compound key groupings from the higher key were used to define the relationships. However, it may be that there is a direct relationship between Installed Machine and Part Supplied since part of the key of the latter contains the composite key of the former. These issues are worth pursuing although in this instance the user cannot identify the need for such a direct relationship.

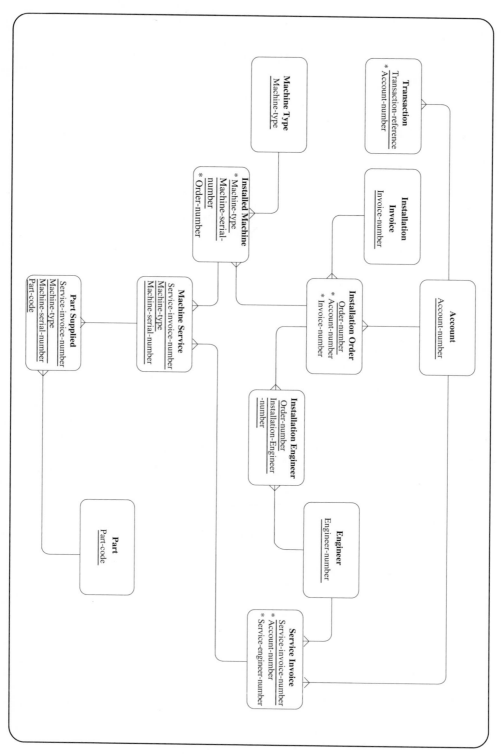

Figure 6.19   The resultant relational data sub-model

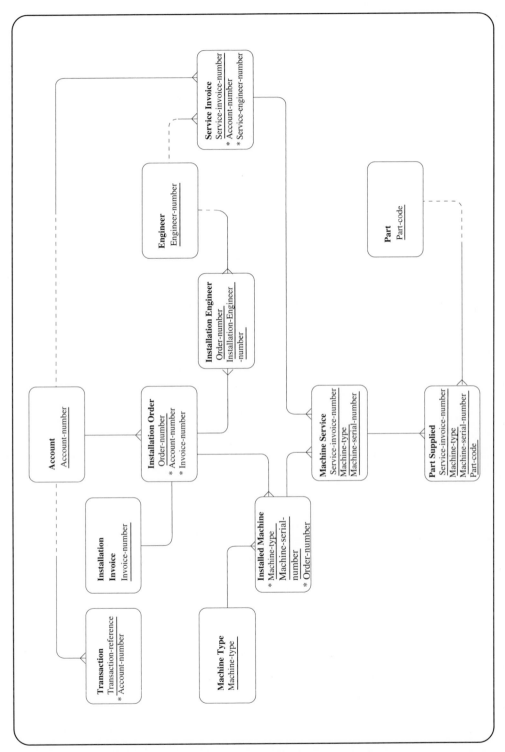

Figure 6.20 Modified relational data sub-model showing 1:1 and optional relationships

6.5.4   Recursive (Involuted) Relationships

Recursive relationships occur when a relation includes in its non-key attributes its own primary key as a foreign key. For example, if the Engineer relation were to include each engineer's supervisor, the relation would be as in Figure 6.21(a) and the corresponding data sub-model as in Figure 6.21(b).

The relationship is optional as not all engineers are supervisors and some supervisors may not themselves be supervised by engineers.

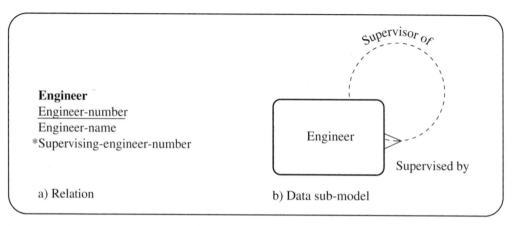

Figure 6.21   Recursive relationship

6.5.5   Identifying Multiple Relationships

Multiple relationships can occur between a relation and itself or between a pair of relations. For example Figure 6.22 illustrates a multiple recursive relationship and Figure 6.23 a multiple relationship between a pair of relations. Multiple relationships between a pair of relations could be represented by the two foreign keys being in one relation.

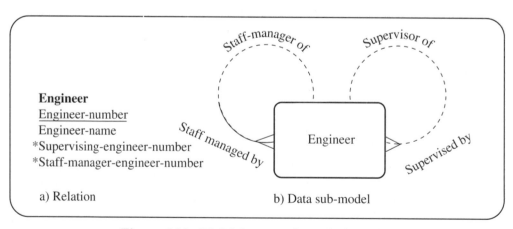

Figure 6.22   Multiple recursive relationships

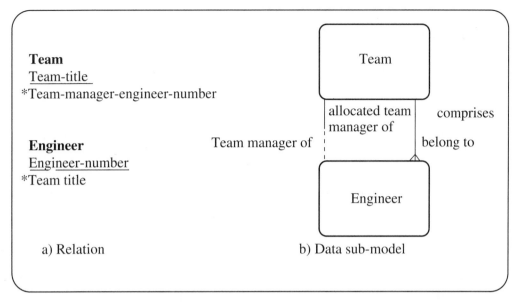

Team
Team-title
*Team-manager-engineer-number

Engineer
Engineer-number
*Team title

a) Relation

Team

allocated team manager of

comprises

Team manager of

belong to

Engineer

b) Data sub-model

Figure 6.23   Multiple relationships

6.6   REPRESENTING AN LDM AS A SET OF RELATIONS

An LDM can be represented as a set of relations, which may or may not be normalised. The purpose of the alternative representation could be to check that part or whole of the LDM is in 3NF. If it is not, then relational data analysis can be performed. Once in 3NF, the relations can once again be represented as a relational data sub-model and merged back into the LDM.

The following guidelines enable the conversion:

- Each entity type becomes a relation.

- The attributes of the entity type become the attributes of the relation.

- The identifier of the entity becomes the primary key of the relation.

- A one-to-many relationship is represented by a foreign key. The primary key of the master entity is copied into the relation representing the detail entity.

- A one-to-one relationship is a special case of a one-to-many relationship. A choice is made as to which entity type is the master and which the detail. If the occurrences of one entity type must be associated with the other entity type, then this would normally become the detail.

- A many-to-many relationships is represented by a new relation consisting only of the attributes that form the primary keys of the entity types. All attributes of the new relation form its primary key.

- It is necessary to check that the primary key of an entity type has not already inherited the primary key of its master entity type. Such attributes may be documented as foreign keys in the Entity Description.

6.7   COMPARING THE LDM AND THE RELATIONAL DATA SUB-MODEL

At any time in the development of the required LDM it is possible to use relational data sub-models to add value to the LDM or simply to validate the LDM. Such 'bottom up' validation will be based on applying relational data analysis to various input sources and I/O Structures. The number of I/O Structures to which RDA is applied will depend on such factors as the complexity, volume, frequency, and importance of such structures.

Once a set of 3NF relations has been represented as a data sub-model it can be compared to the LDM. Care must be taken where the naming of attributes and relations is concerned. Even when strict naming standards have been adopted it is still possible to have the problems of homonyms and synonyms. For example, Order-number may be a homonym for Sales-order-number and Purchase-order-number. Requisition-number may be a synonym (or alias) for Purchase-order-number or be an attribute in its own right. Similar problems arise with entity and relation names. When entity and relation names do not appear to correspond, it will be necessary to match the attributes of one to the other.

The comparison should add value to the LDM. This may be by:

- splitting existing entities into two or more normalised entities;
- identifying alternative primary keys for entities;
- merging entities on the same primary key;
- adding additional attributes to entities;
- adding additional entities;
- adding additional relationships between entities.

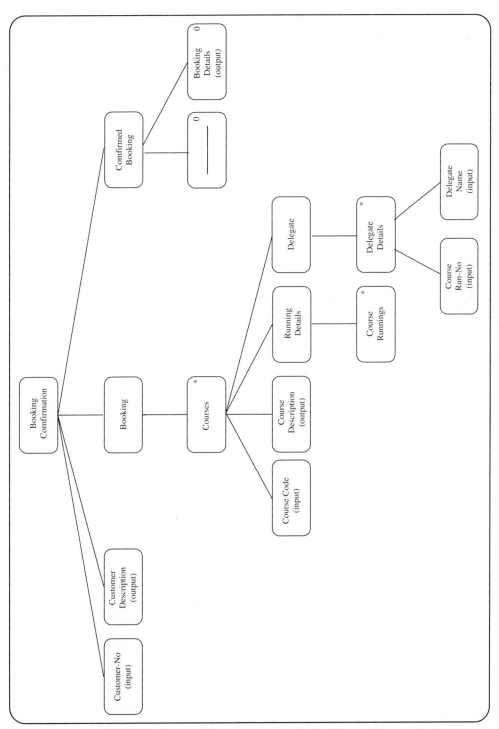

Figure 6.24 I/O Structure Diagram for registration of course bookings

I/O Structure Description					*SSADM* Version 4

Project/System	Author	Date	Version	Status	Page 1
EXAMPLE	SRS	13/2			of 1

Data flows represented   **Established customer bookings**

I/O structure element	Data item	Comments
CUSTOMER NUMBER	CUSTOMER ✳	Bookings only
CUSTOMER DESCRIPTION	CUSTOMER NAME	accepted from
	CUSTOMER ADDRESS	established customers
COURSE CODE	COURSE CODE	Maybe more than one
COURSE DESCRIPTION	COURSE NAME	
	COURSE DURATION	
	COURSE COST	
COURSE RUNNINGS	COURSE RUN ✳	⎫
	COURSE RUN DATE	⎪
	LOCATION	⎬ Repeating group
	ACCOM - FEE	⎪
	PLACES BOOKED	⎭
COURSE RUN - NO	COURSE RUN ✳	
DELEGATE NAMES	DELEGATE NAMES	Maybe more than one
BOOKING DETAILS	BOOK - DATE ⎱	System derived
	BOOK CLERK ⎰	Dialogue confirmed
	BOOKING - REF	System derived

SSADMv4 27

Figure 6.25   I/0 Structure Description for Established Customer Bookings

6.8   REPRESENTING I/O STRUCTURES AS UNNORMALISED DATA LISTS

A single I/O Structure and its associated I/O Structure Description can be the subject of relational data analysis.

Within the I/O Structure:

- Iteration represents a repeating group and corresponds to a nested relation.

- Optional elements are normally considered as separate nested relations. Such nested relations, if they have at most one occurrence, may not have a primary key component of their own. During the process of normalisation they will inherit the primary key of the enclosing relation.

The I/O Structure and I/O Description in Chapter 5 are reproduced in Figure 6.24 and 6.25 respectively.

Figure 6.26 shows the normalisation of the I/O Structure on an RDA Working Paper, and Figure 6.27 the equivalent relational data sub-model.

6.9   ADVANTAGES OF 3NF DATA MODELS

Third Normal Form data models produce simple and non-redundant data structures. This reduces potential inconsistencies in the data. For example, changing the name of an engineer in Service Call-2 relation (Figure 6.12) which is in 2NF requires multiple occurrences to be searched for and replaced. In the corresponding 3NF relations Service Call-3 and Service Engineer-3 (Figure 6.14) the change needs only to be made once. Two further advantages of 3NF relations are concerned with insert and delete anomalies.

6.9.1   Insert Anomalies

A primary key cannot contain a null value. Consequently, details of a newly employed engineer cannot be added to the 1NF or 2NF relations until the the engineer has been involved in either an installation or a service call. However, if the data is in 3NF, the new engineer details can be inserted into the Engineer relation as soon as employment commences.

RDA Working Paper					SSADM Version 4
Project/System	Author	Date	Version	Status	Page of

Source name **Established Customer Bookings**

UNF Attributes	Level	1NF	2NF	3NF	Result Relation	Result Attributes
		<u>Booking reference</u>		<u>Booking reference</u>		<u>Booking reference</u>
Customer no.	1	Customer no.		Booking date		Booking date
Customer name	1	Customer name	⟶	Booking clerk	Booking	Booking clerk
Customer address	1	Customer address		Customer no.	✱	Customer no.
Booking date	1	Booking date				
Booking clerk	1	Booking clerk		<u>Customer no.</u>		<u>Customer no.</u>
				Customer name	Customer	Customer name
<u>Booking reference</u>	1			Customer address		Customer address
<u>Course code</u>	2	<u>Booking reference</u>	<u>Booking reference</u>			
Course name	2	<u>Course code</u>	<u>Course code</u>	⟶		
Course duration	2	Course name				
Course cost	2	Course duration	<u>Course code</u>			<u>Course code</u>
		Course cost	Course name	⟶	Course	Course name
			Course duration			Course duration
			Course cost			Course cost
		<u>Booking reference</u>	<u>Booking reference</u>			
		<u>Course code</u>	<u>Course code</u>	⟶		
<u>Course run no.</u>	3	<u>Course run no.</u>	<u>Course run no.</u>			
Course run date	3	Course run date				
Location	3	Location	<u>Course code</u>			<u>Course code</u>
Accommodation fee	3	Accommodation fee	<u>Course run no.</u>			<u>Course run no.</u>
Places booked	3	Places booked	Course run date		Course run	Course run date
			Location	⟶		Location
			Accommodation fee			Accommodation fee
			Places booked			Places booked
<u>Course run no.</u>	3	<u>Booking reference</u>				
<u>Delegate name</u>	3	<u>Course code</u>	<u>Course code</u>			<u>Course code</u>
		<u>Course run no.</u>	<u>Course run no.</u>	⟶	Delegate booking	<u>Course run no.</u>
		<u>Delegate name</u>	<u>Delegate name</u>			<u>Delegate name</u>
			Booking reference		✱	Booking reference

SSADMv4 40

Figure 6.26   RDA Working Paper for Established Customer Booking

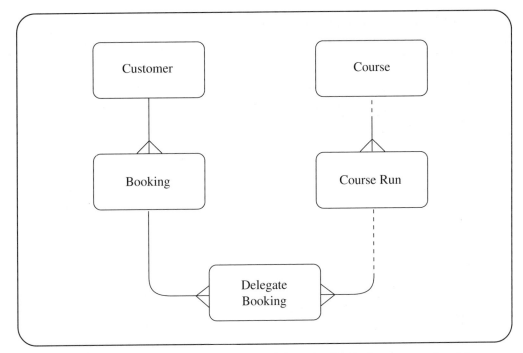

Figure 6.27   Relational data sub-model for established customer bookings

6.9.2   Delete Anomalies

Similarly, if data is not in 3NF, deletions of rows can cause the loss of valid information. For example, a service cancellation causes the relevant row of data to be deleted. In the 1NF relation Service Call or 2NF Service Call-2 relation (Figure 6.13), the deletion of a row would cause the deletion of service engineer details if this is the only work he has been involved in. This is not so for the 3NF relations.

A further benefit of data in 3NF is the ease of extension of the data structure. Thus new information required about existing relations is usually satisfied by adding attributes. Other changes such as recording details of parts available for each Machine Type simply requires an additional relation for Parts Available.

Changes of business rules, for example, the possibility of issuing more than one Invoice for each Installation Order, results in much more minor changes to the specification than a data structure based on process requirements.

In general processing is logically simpler when using 3NF data structures, although this is offset in practice by the performance problems of traversing many relations. However, data in 3NF is more easily shared between applications than data structured for a particular set of application processes. Consequently, new applications or processes are able to access the data as effectively as existing ones.

6.10   FURTHER NORMALISATION

For the vast majority of applications, analysing data to produce 3NF relations is sufficient to ensure that each relation contains no redundancy and hence no update anomalies. However, potential problems may still exist in those relation which have primary keys consisting of multiple attributes. For such relations further normalisation can be performed.

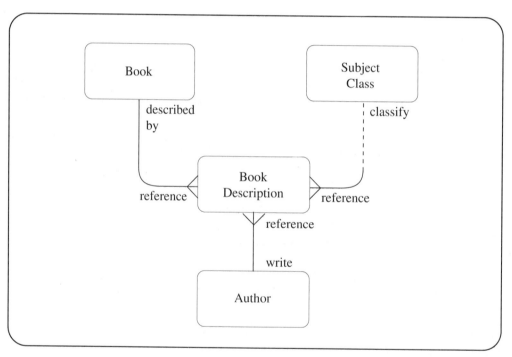

Figure 6.28   Partial LDS in 3NF

6.10.1   Fourth Normal Form (4NF)

Fourth Normal Form (4NF) is concerned with multi-valued dependencies. A functional dependency indicates that for each value of the attribute X there is only one value of the attribute Y. A multi-valued dependency states that for each value of the attribute X there is a finite set of values for Y. If a relation contains two or more independent multi-valued dependencies, additional partitioning to two or more relations will be required.

The entity Book Description in the partial LDS, Figure 6.28, may be represented by the relation illustrated in Figure 6.29.

All attributes in this relation are in the key and there are no functional dependencies between these attributes. However, although in 3NF, this relation still displays redundancy and exhibits update problems. The problems arise because each of the books has two authors and so the title is recorded more than once. This duplication is compounded when a book is referenced under more than one class-title.

It can be seen that

> Class-title is multi-valued dependent on Book-name and Author-name is multi-valued dependent on Book-name.

These multi-valued dependencies may be identified by grouping the attributes in a similar way to the dependency diagrams (Figure 6.30). This figure serves to illustrate that these two multi-valued dependencies are independent of each other. There is no dependency of any sort between Class-title and Author-name.

To remove the redundancy, the relations need to be partitioned into two 4NF relations as illustrated in Figure 6.32 which gives the partial LDS in Figure 6.31.

Book Description		
Book-title	Class-title	Author-name
Introducing Systems Analysis	Systems Analysis	Skidmore
Introducing Systems Analysis	Systems Analysis	Wroe
Introducing Systems Design	Systems Design	Skidmore
Introducing Systems Design	Systems Design	Wroe
SSADM: A Practical Approach	Systems Analysis	Ashworth
SSADM: A Practical Approach	Systems Design	Ashworth
SSADM: A Practical Approach	Systems Analysis	Goodland
SSADM: A Practical Approach	Systems Design	Goodland

Figure 6.29   Book Description relation in 3NF

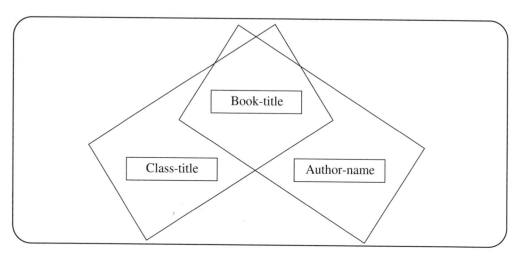

Figure 6.30 Identifying multi-valued dependencies

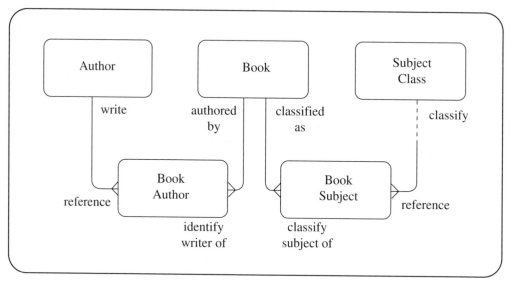

Figure 6.31 Partial LDS in 4NF

Just as before, the partitioning of relations into Fourth Normal Form should not lose information. Thus, it is possible to join the two 4NF relations in Figure 6.32 to give the 3NF relation of Figure 6.29.

Book Author	
Book-title	Author-name
Introducing Systems Analysis	Skidmore
Introducing Systems Analysis	Wroe
Introducing Systems Design	Skidmore
Introducing Systems Design	Wroe
SSADM: A Practical Approach	Ashworth
SSADM: A Practical Approach	Goodland

Book Subject	
Book-title	Class-title
Introducing Systems Analysis	Systems Analysis
Introducing Systems Design	Systems Design
SSADM: A Practical Approach	Systems Analysis
SSADM: A Practical Approach	Systems Design

Figure 6.32   Book Description relation partitioned into 4NF relations

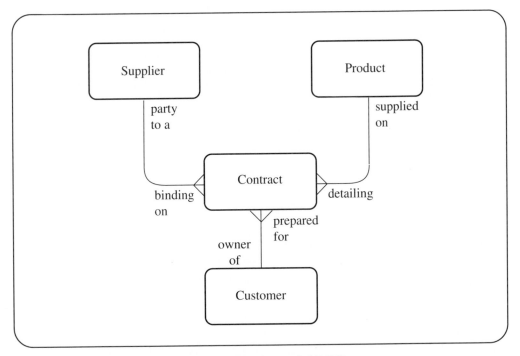

Figure 6.33   A partial LDS

6.10.2   Fifth Normal Form (5NF)

Fifth Normal Form (5NF) is concerned with relations which cannot be partitioned into two relations without losing information, but can be partitioned into three or more relations. The entity Contract in the partial LDS in Figure 6.33 (see previous page) is represented as a relation in Figure 6.34. It is an all key relation which is in 4NF but still exhibits redundancy.

The dependency diagram, Figure 6.35, illustrates that there are multi-valued dependencies but that they are not independent. Moreover, it is understood that the data obeys the join dependency constraint such that:

if

Green supplies Watering cans, and

Watering cans are used by Rose, and

Rose is supplied by Green,

then

Green supplies Watering cans to Rose.

Any attempt at partitioning into two relations would lose information. However, it is possible to partition the relation into three such that if the three relations are joined the original is generated with no loss of information. Moreover, because the data obeys the join dependency constraint there are no update anomalies.

Figure 6.36 shows the partitioned relations and their equivalent LDS is shown in Figure 6.37.

This explanation of Fourth and Fifth Normal Forms is intended only to give a flavour of what these normal forms are about. A more rigorous definition is to be found in other publications (Fagin, 1977, Kent, 1983, Date, 1986). Analysing data to 3NF is more than sufficient for most applications.

Contract		
Supplier-name	Product-name	Customer-name
Green	Watering-can	Rose
Green	Watering-can	Heath
Green	Wheel-barrow	Heath
Brown	Watering-can	Heath

Figure 6.34   The Contract relation

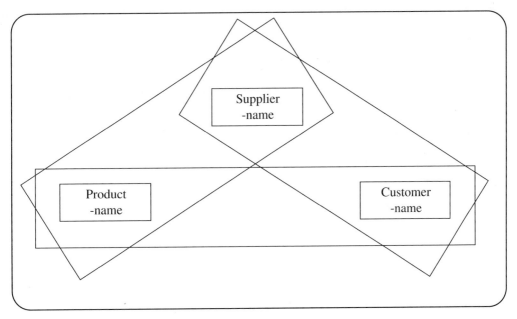

Figure 6.35   Contract 'dependency diagram'

Supplier Product	
Supplier-name	Product-name
Green Green Brown	Watering-can Wheel-barrow Watering-can

Product Customer	
Product-name	Customer-name
Watering-can Wheel-barrow Watering-can	Rose Heath Heath

Customer Supplier	
Customer-name	Supplier-name
Rose Heath Heath	Green Green Brown

Figure 6.36   5NF relations

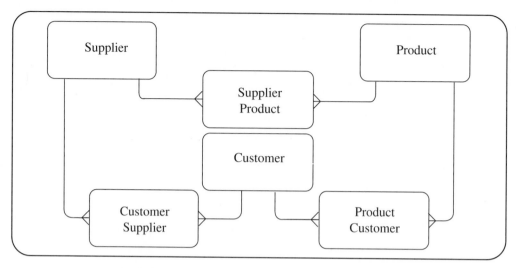

Figure 6.37   LDS Showing 5NF relations

6.11   DOCUMENTATION

The process of Relational Data Analysis may be documented on RDA working papers (Figure 6.26). The dotted lines separating the individual relations are not mandatory but to aid clarity. The results of the process may add to the documentation of the LDM. Each of the attributes should be fully documented on an Attribute/Data Item Description form.

6.12   RELATIONSHIP WITH OTHER MODELS

Relational Data Analysis is a complementary technique to Logical Data Modelling and supports it as a supplementary approach to identifying and specifying information requirements.

Logical Data Modelling is essentially a top-down approach focusing on data that is important to the application area. In contrast, relational data analysis derives a data model from bottom-up and this ensures that all the data details are considered.

The LDM is verified against the Function Descriptions by relational data analysis of appropriate I/O Structures and I/O Structure Descriptions (see Chapter 5). Complex, significant and common inputs and outputs are usually selected for this analysis.

The eventual Required System LDM in Third Normal Form provides the basis for Logical Database Design.

6.13 SUMMARY

The purpose of RDA is to:

- capture the user's detailed knowledge of the meaning and significance of the data;
- validate the LDM to ensure it:
  - is in 3NF;
  - takes account of processing requirements;
  - contains all the required detail.
- ensure the data is logically easy to maintain and extend as all:
  - data inter-dependencies have been identified;
  - ambiguities have been resolved;
  - unnecessary duplication of data (redundancy) has been eliminated.
- form the data into optimum groups to provide a basis for sharing data across many applications.

Some of the steps of RDA appear, to the experienced analyst, to be counter-intuitive or over defined. However, SSADM is designed to produce a consistent, documented approach to analysis and design and hence a standard (some would say cook-book) method of relational data analysis has been defined.

The principles of RDA have not changed from SSADM Version 3 but its application has. Its use in SSADM Version 4 is much more flexible. In Version 3 the LDS and the normalised data model (TNF Data Structure) were produced independently. A Composite Logical Data Design (CLDD) was subsequently developed which combined the strengths of the two data models. It was the CLDD on which physical data design was based. In Version 4, RDA is used as an important and integral part of the process of creating and verifying the 3NF Required System LDM and the Logical Database Design is based on this 3NF Model.

7 Entity-Event Modelling – Entity Life Histories

7.1 INTRODUCTION

So far the modelling of the required system has produced two views:

- The Function Definitions and the Data Flow Model (DFM) which describe the processing in the system.

- The Logical Data Model (LDM) which models the information structure of the system.

Entity-event modelling (EEM) provides a third view of the system requirements. It investigates and models the events which trigger the processes which, in turn, have an effect on the information structures. Entity-event modelling is concerned with identifying events and the sequence in which those events occur. It provides a mechanism by which the data and functional views can be validated against each other to provide a sound basis for design.

Two modelling techniques are used in entity-event modelling:

- Entity Life History (ELH) and

- Effect Correspondence Diagramming (ECD).

The Entity Life History is a major analysis technique within SSADM. It shows the sequence in which events affect each entity and models time and business constraints. It validates the data views of the system and identifies errors and omissions which lead to further detailed processing and data requirements, so producing a complete and consistent specification.

The Effect Correspondence Diagrams are derived from the set of ELHs to show how the different effects caused by events interrelate.

The Effect Correspondence Diagrams (examined in Chapter 8) are subsequently used to specify the Update Process Models in Logical Design.

7.2 MODELLING NOTATION AND CONCEPTS

In order to construct ELHs it is necessary to understand the fundamental concepts on which they are based.

7.2.1   Event

An ELH shows the events which may have an effect on a particular entity occurrence. An event can be defined as something that triggers a process to update system data. An event is not a process, it is the *stimulus* which causes that process to be invoked.

Each event must have a unique name, which should reflect what is causing the process to be invoked, for example, Receipt of Examination Result and Notification of Examination. In practice, due to the diagram space constraints, event names are frequently shortened: Examination Result, Examination Notification, Resignation etc . . . Care should be taken that the name reflects the event rather than the DFD process, otherwise there is a danger that other events triggering the same process may be missed.

Three types of event can be distinguished:

– external event	– a transaction arriving from the outside world;
– internal process event	– occurs when a predefined condition within the stored data has been met;
– time-based event	– takes place when a particular process is to be triggered at a regular time interval: a set time of day, month, year, etc.

A single event instance may cause more than one entity occurrence to change. The changes within a single entity occurrence caused by an event is called an effect.

7.2.2   Effect

A single event will cause a change to at least one entity occurrence of at least one entity or to its relationships with other entity occurrences. Each such change is called an effect.

An effect can be:

- creation of a new entity occurrence;

- deletion of an existing entity occurrence;

- modification of existing entity occurrence.

This last category, modification effects, can be:

- storing an attribute value;

- modifying an attribute value;

- changing relationships.

The effects on relationships include:

- tying the entity occurrence to a master entity occurrence;

- cutting the entity occurrence from its master entity occurrence.

It is valid for one event to affect an entity occurrence at more than one point within its life or, at different times, to invoke substantially different processing.

7.2.3   Mutually Exclusive Effects

There may be occasions when a single instance of an event may affect an entity in one of several mutually exclusive ways. For example, the event Transaction acting on the entity Account might have different effects when the transaction is a payment from when the transaction is a withdrawal. A withdrawal transaction might require that a check on the balance of the account is performed, and a count of the number of withdrawals or the value of withdrawals within a specified time period is maintained. A payment transaction might only update the account balance and record transaction details. In these circumstances, the event name should be qualified by a description of the exclusivity, for example, Transaction(payment) and Transaction(withdrawal). The processing required will be determined at the time of the event either by the stored data or by the data accompanying the event.

7.2.4   Entity Roles

If a single event affects more than one occurrence of a particular entity, and the effects are different for each entity occurrence, then the entity is deemed to be assuming different roles. As processing will need to be specified for each role, each different role must be separately identified on the ELH for the relevant entity. On the ELH the event name must be qualified by the role which the entity is assuming. For example, if the business rules are such that a patient can be registered with only one doctor at any one time, notification of doctor registration would have the effect that the old registration must be cancelled and a new registration created. In this case there are two roles for the one event, Notification of Registration [old] which cancels the current registration, and Notification of Registration [new] which creates a new registration occurrence.

The role names are used to distinguish between the different effects of the event. Note that different brackets are used for role qualification and for mutually exclusive effects.

7.2.5   Notation of the ELH

An ELH is drawn as a tree structure and uses the same basic notation as the SSADM Structure Diagrams:

- nodes are drawn as square-cornered boxes;
- the root node represents an entity type and contains the name of the entity;
- the elementary (leaf) nodes represent events which have an effect on the life of an entity occurrence of the entity type;
- the elementary nodes (effects) contain the name of the event.

The typical life of an entity starts with an event which triggers processing to create a new entity occurrence. Once an entity occurrence has been created, then an event can trigger processing whose effect is to modify attribute values of that occurrence. At some later time in the life of the entity occurrence, an event will occur which will trigger processing which will have the effect of terminating the life of that occurrence. A terminating event simply means that the occurrence is no longer of interest to the system. This may mean that the occurrence is archived or transferred to another file for use by other systems. This typical life of an entity illustrates the fundamental structure of an ELH.

7.2.6   Sequence

The sequence structure is fundamental to all ELHs. This structure is shown in Figure 7.1. It shows the chronology of events for a Course Attendance. Notification of Enrolment (create occurrence event) always takes place before the Written Examination Result can be notified. The Written Examination Result event occurs before the Notification of Graduation (terminate occurrence event).

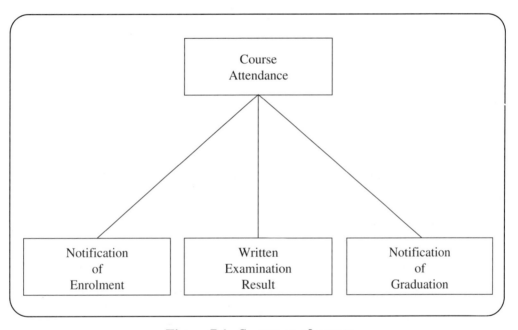

Figure 7.1   Sequence of events

7.2.7   Selection

In Figure 7.2 the small circles in the two Assessment events denotes that these are alternate events. Thus an assessment event is either a notification of a Written Examination Result or a notification of an Oral Examination Result. The alternative events are grouped under a selection node.

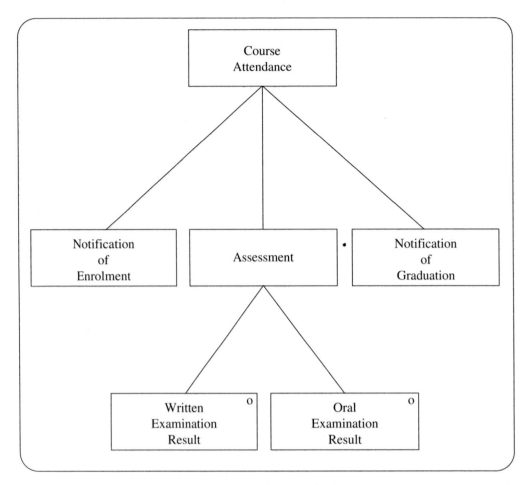

Figure 7.2 Selection of events

Figure 7.3 illustrates the use of the null event which enables the possibility of neither event having an effect on a particular entity occurrence. Either notification of a Written Examination Result or notification of an Oral Examination Result can happen following Notification of Enrolment and either will be followed by the Notification of Graduation event. Alternatively, the Notification of Graduation can follow Notification of Enrolment because of the inclusion of the null event in the selection.

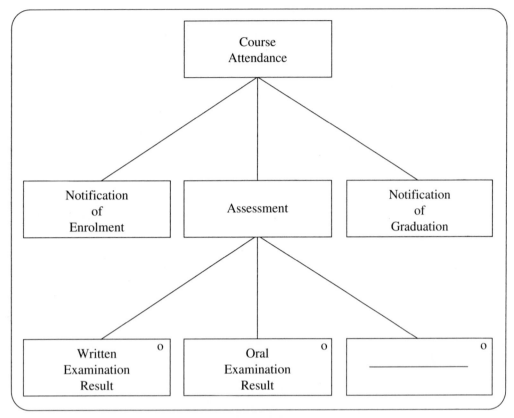

Figure 7.3 Null event

7.2.8 Iteration

The asterisk in the effect box in Figure 7.4 denotes that the event may affect an entity occurrence zero or more times. Continuous Assessment Result may have zero, one or many Results before the Notification of Graduation occurs.

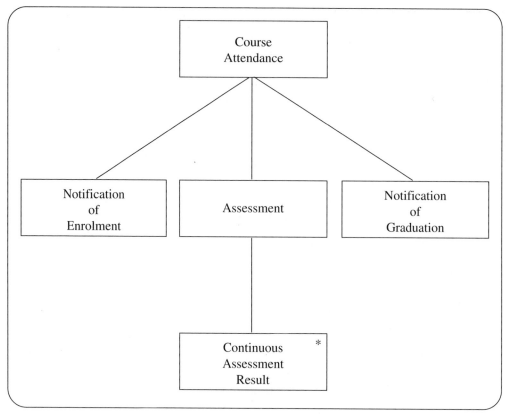

Figure 7.4 Iteration of an event

Figure 7.5 illustrates the iteration of a substructure of the ELH. In this case the sequence of events notification of Written Examination Result and notification of Oral Examination Result may not occur, may occur once, or may occur many times. Each instance of the iterated sequence must be complete before the next sequence can begin. Thus the business rule is recorded that a subsequent written examination cannot take place until an oral examination has occurred, or more usually, every written examination is followed by an oral examination.

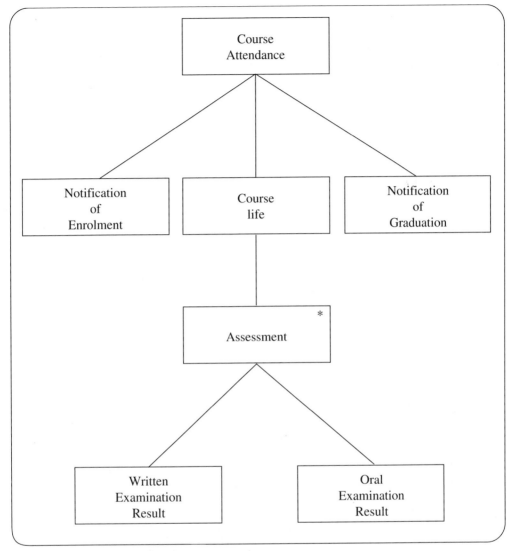

Figure 7.5   Iteration of a sequence

7.3   CONSTRUCTION OF ENTITY LIFE HISTORIES

The ELH technique is used to investigate the lives of each entity, identifying the events which have an effect on their lives, documenting the way in which the lives are affected and showing the sequence in which the effects take place.

Thus an ELH represents all the permitted sequences of events that may occur during the life of an occurrence of an entity. Each occurrence is constrained to behave in a way defined by the ELH for that entity. However, at any one time each entity occurrence may be at a different stage in its life history, and each entity occurrence may lead permitted lives that are different from those of other occurrences of the same type. An ELH is a description of all possible lives for every occurrence of the entity.

The basic steps in ELH construction are summarised below:

- create the Event/Entity Matrix;
- draw initial Entity Life Histories;
- review Entity Life Histories;
- add operations;
- add state indicators.

To illustrate these steps, a small part of a Public Library lending system will be developed in three stages. The LDS for the system is given in Figure 7.6.

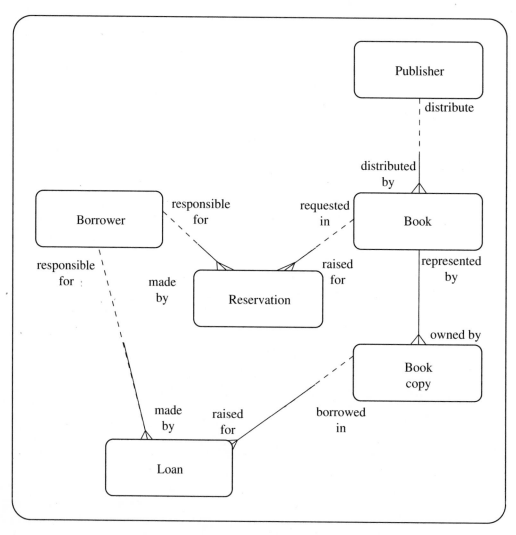

Figure 7.6 LDS for a public library system

The public library maintains a catalogue of books. Entries to this catalogue are made when a copy of a book has been acquired. Before copies of books are available for loan they must be classified using the Library standards. Only registered borrowers may loan copies of books. When borrowers reserve a book which is not available, a reservation is placed on all copies of the book. When the first copy is returned, it is allocated to the reservation and the other copies of the book released from the reservation. All Loan and Reservation records are retained until the end of year when reports are produced analysing the duration and quantity of loans, classification of books reserved, etc. Borrower details are not required for this analysis.

7.3.1   Create the Event/Entity Matrix

The Event/Entity Matrix is produced as a working document to aid the drawing and checking of the Entity Life Histories. It is not used as an input to any other model and may be disposed of once the Entity-Event modelling is complete.

The Event/Entity Matrix is a two dimensional grid for recording which entities each event affects and the nature of that affect. It initially provides two simple checks:

– that each entity is affected by at least one event;

– that each event affects at least one entity.

The columns of the matrix are labelled with the Entities from the LDS and the rows of the matrix are the events.

An external event may be recognised on the DFD as the arrival of a data flow from an external source. An internal process or time-based event may be recognised by a process with input data flows from data stores only.

For each event it is necessary to identify which entities are affected. The Logical Data Store/Entity Cross Reference shows which entities are represented in each data store and the data items on the input data flows can be compared with the Entity Descriptions to identify which attributes are being affected and the nature of that effect.

An effect can can be classed as:

– creation of an entity occurrence;

– modification of attribute values in an entity occurrence;

– deletion of an entity occurrence.

C (creation), M (modify) or D (deletion) is entered into the cell forming the intersection of the event row and the entity column. Some events may have effects on more than one entity, while other events may have a combination of effects on the same entity according to role. Figure 7.7 shows the exploratory matrix for the library system.

Once the matrix has been completed, it can be reviewed for completeness. For each column, that is each entity, there should normally be at least one creation (C) effect and at least one deletion (D) effect, unless it has been specified that occurrences of this entity are never deleted. If not all attribute values of the entity are to be stored at creation time,

Event/Entity Matrix									*SSADM* Version 4							
Project/System				Author			Date		Version		Status		Page of			

Event \ Entity	Book	Book Copy	Publisher	Loan	Borrower	Reservation										
Notification of Acquisition	C	C														
Registration of Borrower					C											
Issuing of Loan		M		C												
Return of Loan		M		M	M											
End of year Analysis			M													
Classification of copy		M														
Notification of Disposal		D														
Notification of Resignation					D											
Request for Reservation		M				C										
Renewal of Loan				M												

SSADMv4 20

Figure 7.7   Exploratory Event/Entity Matrix for Library System

then there should be at least one modify (M) effect which establishes these values. If established attribute values are changed during the life of the occurrence this should also be shown by modify (M) effects.

In Figure 7.7, the Event/Entity Matrix, it can be seen that a possible life for a Book Copy consists of creation by the Notification of Acquisition event, modification by the Issuing of Loan event (On-loan-flag is set), modification by the Return of Loan event (On-loan-flag is reset), modification when classified and modification by a reservation (Reservation-flag set). The Book Copy occurrence is deleted on Notification of Disposal. The Loan entity is created when a book is issued, modified when the loan is returned or renewed, and deleted following the end of year analysis.

Reviewing the matrix for completion it can be seen that:

 – the Book entity occurrences are created but not modified or deleted;

 – the Borrower entity occurrences are created and deleted but not modified;

 – the Reservation entity occurrences are not deleted;

 – the Publisher entity has no events affecting it at all.

These issues and anomalies need to be resolved by further analysis and by discussion with users.

Compiling the Event/Entity Matrix is an optional activity within SSADM. However, it can be seen that it provides an initial record of the effect of each event on the entities as well as being an important cross-checking mechanism. We believe it is an important first step in the successful construction of ELHs.

7.3.2   Draw the Initial Entity Life Histories

The Event/Entity Matrix does not completely represent the life of an entity occurrence. It does not show which events may occur repeatedly, which are alternate events, or the sequencing of those events on a single entity occurrence. Furthermore, the different effects depending on the state of the entity occurrences are also not shown. For this the graphical notation of the Entity Life History is required.

One Entity Life History is drawn for each entity on the Required System LDS. Initially only a simple life history is drawn, based on the premise that all occurrences must be created, will probably be modified and will finally be deleted. There will be no consideration of abnormal or exceptional events.

The preferred sequence of completion is to start with the lowest level entities on the Required LDS, that is, entities which are only details, and as each ELH is completed, systematically work on their masters up through the LDS until the top-most master entities are reached. This method of working leads to a better understanding of the life of each entity and the effect on its master(s).

The construction begins by selecting from the Event/Entity Matrix all the events which may cause an occurrence of an entity to be created. If there is more than one event, then these events are shown as selections under a structure box. It is the creation or birth event which stores a value for the primary key and for many of the other attributes.

All the events which cause modification to an existing entity occurrence are now considered. The sequence in which these events will be notified is now decided. Consideration is also given to whether any events occur more than once (iteration), are alternatives (selection) or are optional. A structure is built up using the sequence, selection, and iteration constructs.

Finally the deletion event is added. If there is more than one way that an occurrence can be deleted, then a selection structure should be used.

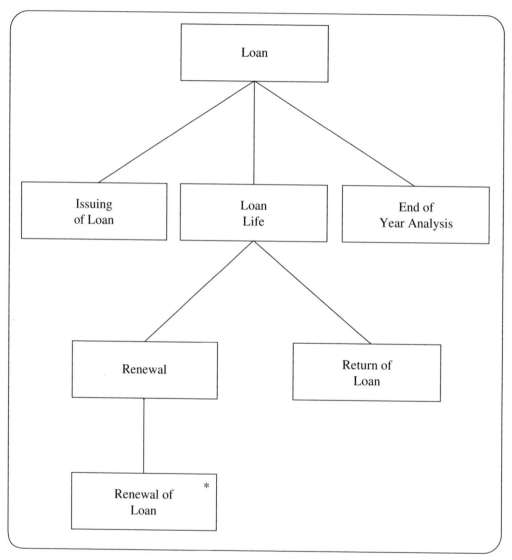

Figure 7.8   Loan ELH

Figure 7.8, showing the initial Loan ELH, illustrates a straightforward life of a library loan. The Issuing of a Loan causes the creation of a loan record which will include a

Date-of-loan and Due-date for return. The optional Renewal of Loan causes the Due-date to be modified, possibly many times. The Return of Loan sets the Return-flag. At the End of Year (a temporal event) all Loans with Return-flag set will be analysed for length of loan and then deleted.

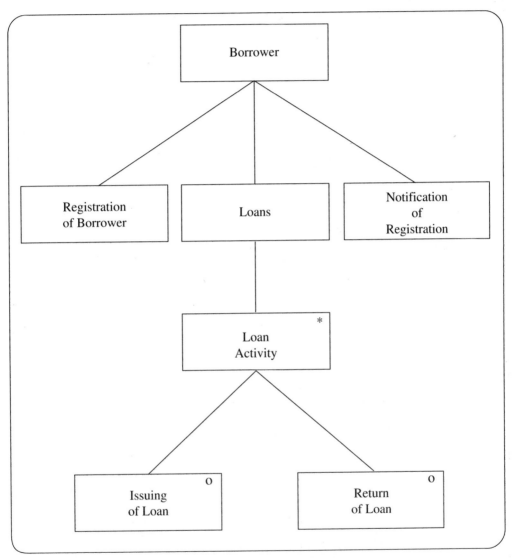

Figure 7.9   Borrower ELH

In the Borrower ELH, Figure 7.9, the effect of the detail on the master can be seen in that the number of loans a borrower currently has is recorded by the Current-loan-count. Thus, Issuing of Loan and Return of Loan modify the Current-loan-count.

In the Loan ELH there is a defined sequence to the Issuing of a Loan and the Return of a Loan events. It is inappropriate to define this sequence in the Borrower ELH, as a borrower may have a number of loans issued to him before returning any of the loans. Moreover, the sequence of the return of loans may be different from the sequence in which the loans were issued.

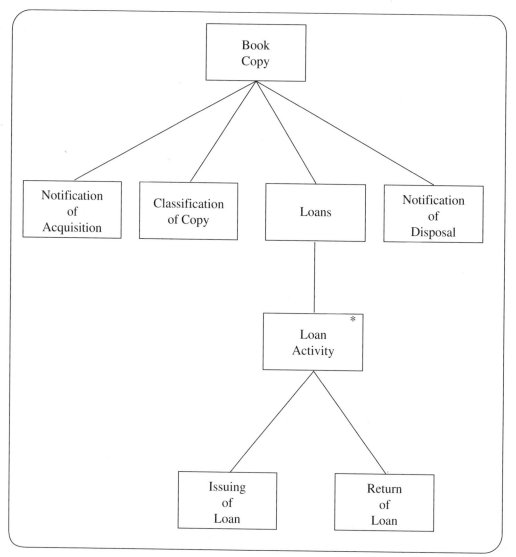

Figure 7.10   Book Copy ELH

Issuing of Loan and Return of Loan also have an effect on Book Copy, the other master entity of Loan. In Figure 7.10 this sequence of events sets and resets the On-loan-flag, and the former event increments the Loan-count which records the number of times a book has been out on loan. The sequence of these events is important, as a copy of a book cannot be loaned until it has been returned.

The ELH for Reservation in Figure 7.11 indicates that once the loan of a Book Copy is returned then the borrower can be notified that the reservation is available for collection. When the borrower responds the reserved book copy can be issued on a loan, which causes the end of reservation state. Alternatively, if the borrower does not respond within three weeks, the reservation will automatically end and the Book Copy made available for further loans. If there is another reservation for the Book this Book Copy now becomes available to fulfil that reservation. Thus, the 3-week non-collection event affects more than one occurrence of the Reservation entity. End of Reservation and Available for Reservation are entity roles for this event. The actual deletion of a reservation is not until the end of year following the analysis of all reservations.

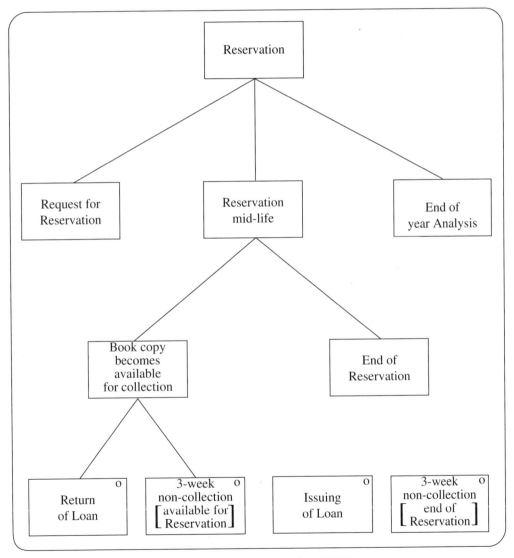

Figure 7.11   Reservation ELH

The ELH for the Book entity, Figure 7.12 is very simple, as are ELHs for many master entities. Note that the death event has been identified with the users – a rarely occurring catalogue review. This would be added to the Event/Entity Matrix if this document is to be retained for future reference.

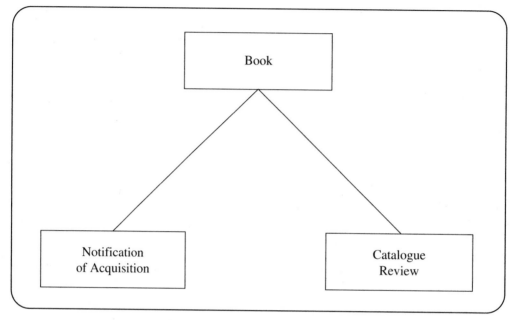

Figure 7.12   Book ELH

Once the simple or normal lives of the entities have been developed it is necessary to consider more complex situations, especially those caused by events which deviate from the expected norm or for which the sequence cannot be predicted. However, before the initial ELHs can be completed some additional notation is required.

7.4   FURTHER NOTATION

7.4.1   Parallel structures

A parallel structure indicates that the sequence of the events cannot be defined. For example, after a house has been placed on the market and valued, and before the sale contract is signed, a number of events may take place associated with prospective sales. However, following any of these events the vendor may change the offer price of the house. The timing of the event cannot be predicted and it does not affect the sequencing of the prospective sale events. This is represented, as in Figure 7.13, by placing the unpredictable event and the group of events which it may precede or follow as two legs beneath the parallel construct.

7.4.2   Quit and Resume notation

A Quit (Q) and Resume (R) is used either when an entity occurrence reverts to an earlier stage in its life or jumps to a later stage. Quits and resumes are linked by a numeric identifier, and there may be many quits associated with one resume.

Figure 7.13 illustrates the use of quit and resume. It shows that it is possible that following a Purchase Offer for a house the next event may be the Signing of Contracts. This is indicated by the Quit (Q1) and the Resume (R1) notation. However, the normal sequence of events is for the Purchase Offer to be followed by the Building Society Survey Valuation.

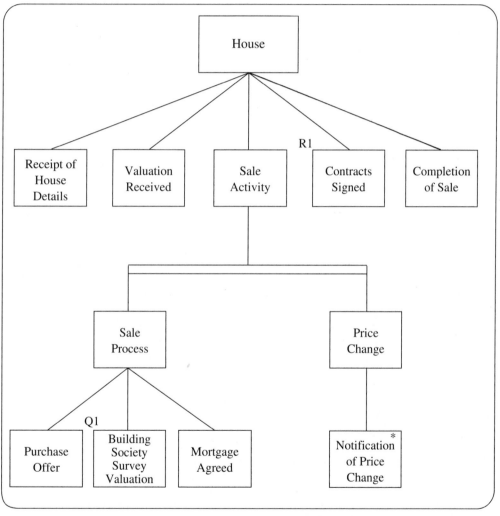

Figure 7.13   Quit and Resume and parallel events notation

7.4.3   Quit from Anywhere Construct

The quit and resume construct can be extended to handle random events. These are events whose timing cannot be predicted, may occur at any point in an entity's life, and may cause abnormal effects or even the termination of an entity occurrence. In this case 'Quit from anywhere to Rn' is written at the bottom of the diagram. The resume event, labelled Rn, is either within the existing structure (Figure 7.14a) or an event which is separate from the main structure in what is termed a sub-structure (Figure 7.14b). Such a sub-structure may consist of a single event or a number of events following the ELH structure rules. If, once the unpredictable event and any subsequent events have occurred, there is a need to revert back to the main structure then the quit and resume notation is used once again.

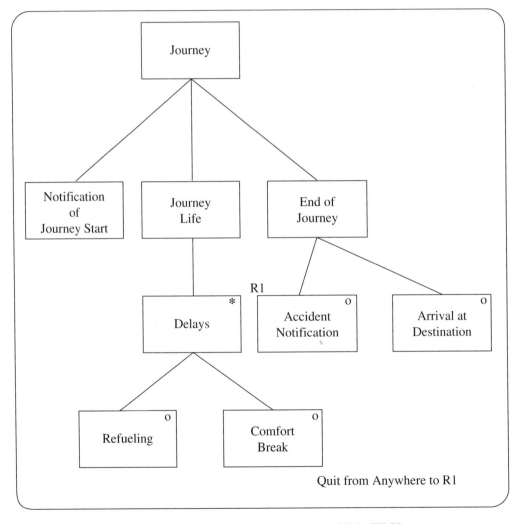

Figure 7.14a   Quit from anywhere to event within ELH structure

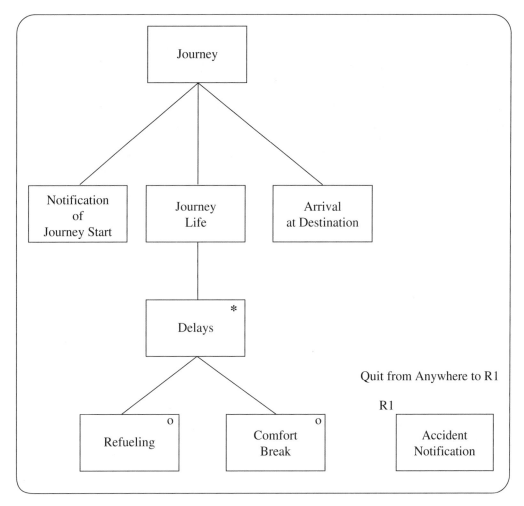

Figure 7.14 b   Quit from Anywhere Notation

Quits and resumes are most powerful when dealing with abnormal or error situations. They should not be seen as a way to cope with complex ELHs and should only be used when no other structure or structures can achieve the requirements. In Figure 7.13 the same business rules could have been defined using a Null event in a selection with the Survey Valuation and Mortgage Agreement events rather than the Quit and Resume shown.

7.4.4   Structure Nodes

Structures may be combined to represent all the possible lives of entity occurrences as long as the event box remains the elementary node of the structure. This grouping under structure nodes ensures that the different component types are not mixed at the same level within one branch of the structure. Whenever possible, structure nodes should be meaningfully named with a group heading which normally defines a type of sub-life or timeframe in which the events occur.

7.5   CONSTRUCTION OF ELHS – COMPLETING THE INITIAL ELHS

The view of library loans has been a very naive one; in reality many loans are not returned by the Due-date. The library rules are that if a book becomes overdue it must be returned to the library, it cannot be renewed.

Moreover, if a loan becomes overdue it is recalled. Each time a loan is recalled the Recall-count is incremented. A loan may need to be recalled a number of times before the Borrower responds by either returning the loan or paying the cost of the book copy they are unable to return. The End of Year analysis will identify the book as lost or returned and logically delete the loan record. The initial ELH for Loan is given in Figure 7.15. Note that the recall event is the temporal event, Over-due Date, rather than 'recall loan' which is the effect of the event.

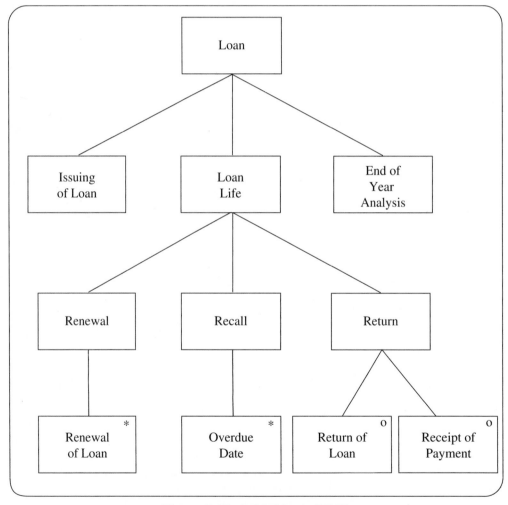

Figure 7.15   Initial Loan ELH

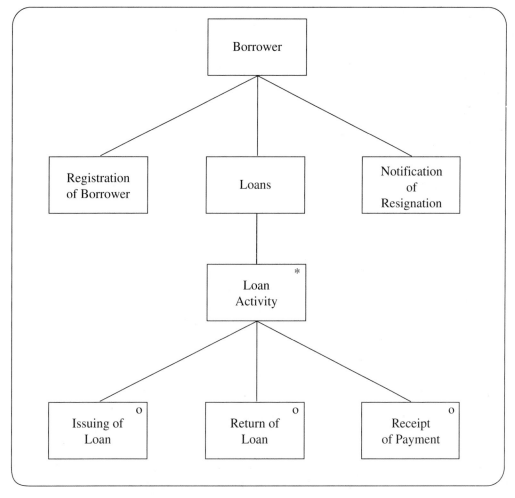

Figure 7.16   Initial Borrower ELH

Receipt of Payment also becomes an alternative event, maintaining the Current-loan-count in the Borrower ELH (Figure 7.16).

Further analysis of the life of the Book Copy entity shows it is also more complex than at first thought.

There are occasions when the Issuing of Loan may unusually be followed by Receipt of Payment because the borrower admits that the copy of the book has been lost. The Quit and Resume notation has been used to define this abnormal sequence of events. Receipt of Payment is included in the Disposal branch of the ELH because it causes the deletion of the entity occurrence. Placing it within the Loan Activity would be incorrect. The normal sequence is that the Return of Loan event occurs and the Book Copy may be the subject of a future loan.

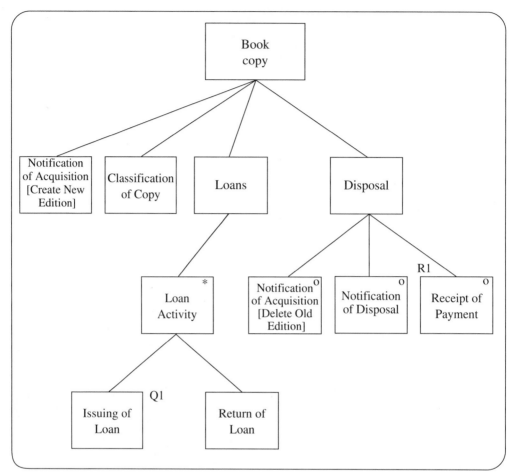

Figure 7.17   Revised Book Copy ELH

Further analysis has also identified another business rule: if a new edition of a book is acquired, all previous editions are withdrawn from the loan stock of the library. Thus the Notification of Acquisition event has two entity roles, deletion of the old edition and creation of a new edition. The revised Book Copy ELH is given in Figure 7.17.

The Book Copy ELH is not complete until the reservation events are considered. The effect of a reservation can happen at any time in the mid-life of a Book Copy. This is modelled using the parallel construct with the loan events on the primary leg and the reservation events on the subsidiary leg. When a reservation is notified, the Reserved-flag is set on all copies of the reserved book. It is reset when the copy is issued on loan to the borrower who reserved it, as indicated on the Reservation ELH (Figure 7.11). It may also be reset, by Release from Reservation, while the Book Copy is still out on loan because another Book Copy occurrence of the reserved book has been returned and allocated to the last reservation for that book. Alternatively, the reservation will end if the reserved copy has not been collected within the three weeks allowed. The initial ELH for Book Copy is given in Figure 7.18.

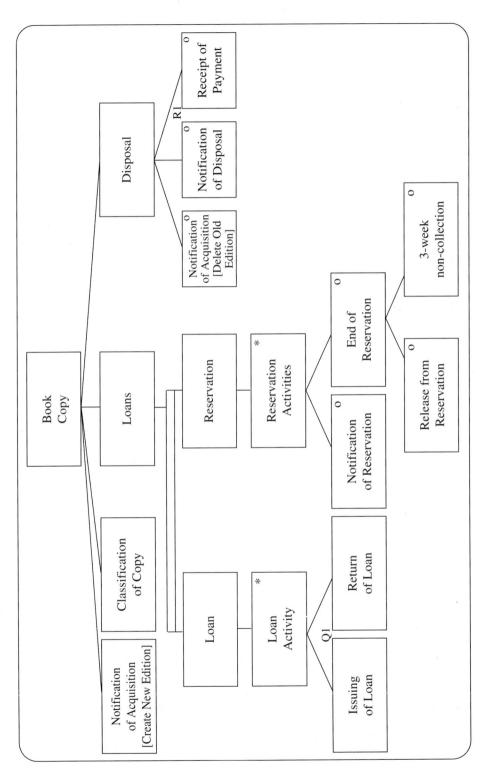

Figure 7.18 Initial Book Copy ELH

ELH analysis is an iterative activity during which the analyst models the business rules using the ELH notation. Anomalies in the current business rules may be identified. For example, as long as the Loan is renewed before the Due-date the Borrower can continue to retain the copy, even though that copy may have been marked as required for a reservation. Resolution of such issues can only be made by the user and by a more detailed analysis of the business procedures.

7.5.1   Review the Entity Life Histories

Each ELH is reviewed, starting with the ELHs of those entities which have no master entities and continuing down the hierarchy of the Required System LDS. The review should highlight all the exceptional events which may cause alterations to the sequencing of events or the progress in the lives of entity occurrences. If the progress of the life of an entity occurrence is constrained by effects in the ELH of another entity, then that constraint must be reflected in the ELH of the entity being constrained.

The following must be considered:

Constraints on a detail

– Death of a master entity will mean the death of its associated detail entity occurrences or transfer of those details to another master entity occurrence. This will not be necessary if the detail has an optional relationship with its master.

Constraint on the master

– Death of a master cannot take place until all of its associated detail entity occurrences have completed their lives normally.

– Following the creation of a master, it cannot progress its life until at least one or possibly all of its associated detail occurrences are created.

Interactions between entities

– An event in the life of an entity may have an effect on its associated master or detail entities.

Reversions

– An entity may revert back to a previous point in its life necessitating the use of the quit and resume notation.

Random events

– Random events may occur at any stage in the entity's life. These will require the Quit from anywhere to Rn where the resume, Rn, is either within the structure or is in a separate event or sub-structure detached from the main ELH.

Non-updating effects of events

– Non-updating processing associated with events which do not have an effect on data but do have an effect on the progress of an entity life have to be considered. For example, there may be significant retrieval events and temporal events which allow the life of the entity occurrence to continue only after they have taken place.

Reviewing the library system ELHs using this checklist reveals that the ELH analysis is not yet complete.

Since a Loan cannot be deleted until the End of Year analysis, the death of a Book Copy occurrence must be constrained by its detail Loan entities. A Book Copy cannot be deleted until all the Loans associated with it have been deleted and this is not until the End of Year analysis has taken place. The corrected ELH for Book Copy (Figure 7.21) includes the End of Year Analysis event as the Book Copy death event; the disposal events now only mark the entity occurrences for deletion.

Book occurrences cannot be deleted, only marked for deletion by the Catalogue Review. If the Book has been the subject of a reservation, the Book information is required in the End of Year Analysis. Consequently, Book is constrained by its Reservation detail until the End of Year Analysis has taken place, after which Reservation occurrences and Book information of disposed of Book Copies can be deleted (Figure 7.23)

Resignation of a Borrower cannot take place until all the borrower's loans have completed their lives, that is, been returned or accepted as lost and a payment of the book's value received. Thus an additional selection to clear the final outstanding loan has been inserted before the Notification of Resignation event in the Borrower ELH (Figure 7.20). The effect of these alternative events includes setting the state of the entity to allow the resignation event to apply.

Reservations may be cancelled. The cancellation may happen at any point in the life of a reservation. Such a random event is modelled using the Quit from anywhere to Rn notation, where the resume event is an off the structure event. To include cancellations in the End of Year analysis it is necessary to return to the main structure using another quit and resume (Figure 7.22). Cancellation of Reservation also needs to be included in the Book Copy ELH as an additional way to end a reservation (Figure 7.21).

It is always possible that a loan is not returned following the last recall. The Library business rules assume that if a book is not returned by six months after the third recall date it is defined as lost in the End of Year Analysis. An additional temporal, non-updating event enables this business rule in the Loan ELH (Figure 7.19). This event also needs to be included in Book Copy as an alternative loss of book event to enable deletion of the book copy which is subject to the non-returned loan (Figure 7.21).

Figures 7.19, 7.20, 7.21, 7.22 and 7.23 show the ELHs for Loan, Borrower, Book Copy, Reservation and Book following this review. However, it must be noted that this ELH analysis is not complete and that not all events associated with a Public Library System have yet been identified. Further events may be discovered by completing or reviewing the business operations.

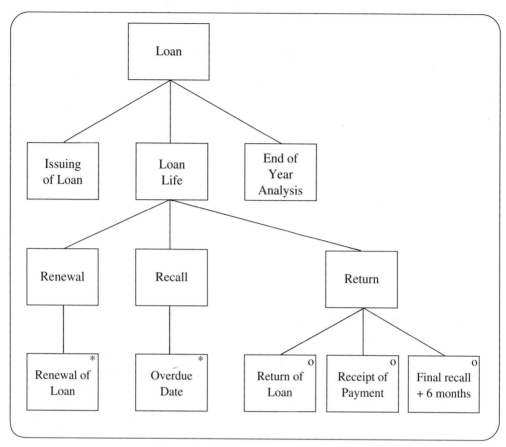

Figure 7.19   Reviewed Loan ELH

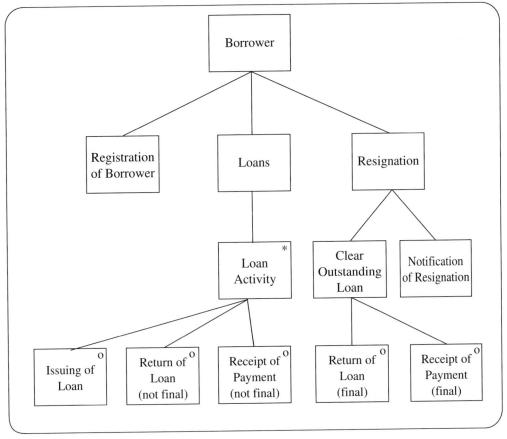

Figure 7.20   Reviewed Borrower ELH

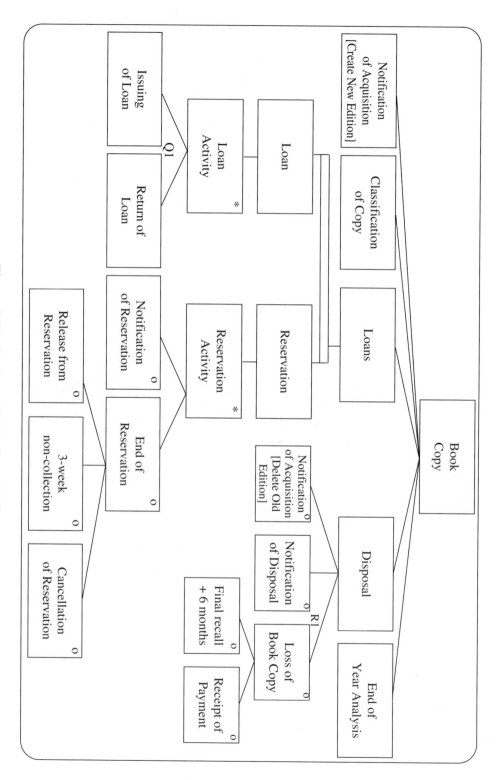

Figure 7.21 Reviewed Book Copy ELH

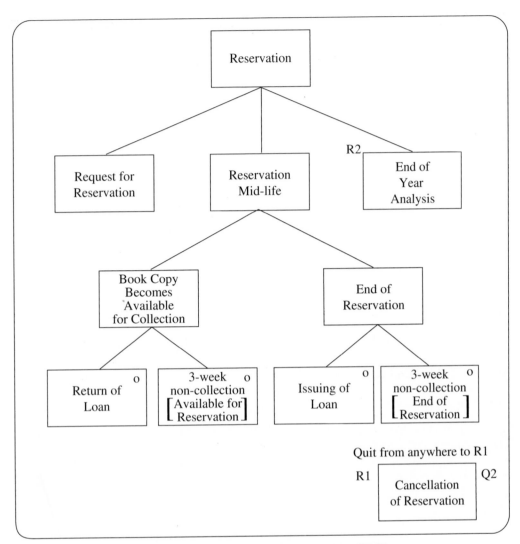

Figure 7.22 Reviewed Reservation ELH

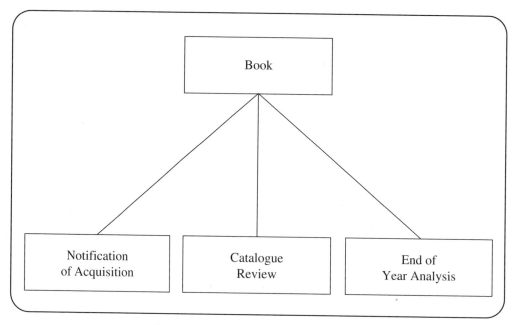

Figure 7.23   Reviewed Book ELH

7.6   MORE NOTATION

7.6.1   Operations

Operations within the ELH technique represent the discrete components of processing which, when combined, constitute the event's effect. For example, the event Receipt of Payment might have the operations Modify Balance-owed and Compute and Store Interest-payable-amount.

Operations can be useful in checking that different events have not been combined. For example, the Receipt of Payment may have a different effect on the entity occurrence depending on whether it is the first payment, a subsequent payment or the final payment. Distinguishing between these different effects may highlight the allowable progress to a subsequent event. The event Closure of Account can only occur after the Receipt of Payment (final) has occurred and not after any Receipt of Payment event.

Operations are useful in validating the LDM by asking which events store each attribute value, which amend those values, and which change relationships. Where attributes are not given a value or not amended there may be a missing event or the attribute may be superfluous to the system requirements. In fact, it is often useful to start to list the operations which may be performed on the entity, referencing them to events, early in the ELH analysis.

7.6.2   Operation Boxes

For each ELH the major operations relating to effects are listed on the ELH or on a separate document and allocated a number. These operation numbers are shown on the diagram in small boxes attached beneath the effect (elementary node) to which they relate. An effect may be the result of more than one major operation; some effects may have no major operations.

7.6.3   State Indicators

State indicators are a method of controlling the sequencing of events. They may be thought of as an additional attribute within each entity. Each time an event affects an entity the state indicator is updated to indicate that the particular effect has occurred. Holding a state indicator in each entity means that it is possible to detect the state a particular entity occurrence has reached within its life. Without a state indicator, complex logic may be needed to investigate many attribute values in different entities.

Knowing the current state of an entity occurrence means that it can be determined whether it is valid to apply the effect of a particular event to an entity occurrence or whether an error condition exists. This implicit validation logic will be carried forward into logical design and built into the processing logic. State indicators also enable selection of entities which have reached a particular stage of their life and so will aid query and report processing. Given that state indicators reflect the structure of the ELH diagrams, their addition to the diagrams is essentially a mechanistic procedure.

7.6.4   Recording State Indicators

For each event the notation used for state indicators is in the form:

valid prior value(s) / set to value

- valid prior value(s) gives the values of the state indicator that must exist for the effect of an event to take place;

- set to value is the value given to the state indicator once the effect of the event has been completed;

- there may be more than one valid prior value of the state indicator for an event;

- there will be only one set to value for each event within each ELH.

Set to numbers are written below each effect box on the diagram. The simplest method is to allocate 1 to the birth effect and to number sequentially through the diagram. The termination event or events do not need a set to state as the entity occurrence will no longer exist in the system for events to affect it. This is indicated by the null state indicator, a hyphen (-).

Prior values are assigned by recording the state value of all effects which may immediately precede the event. The birth event, or events, will not have any preceding states; this is also indicated by the null state indicator.

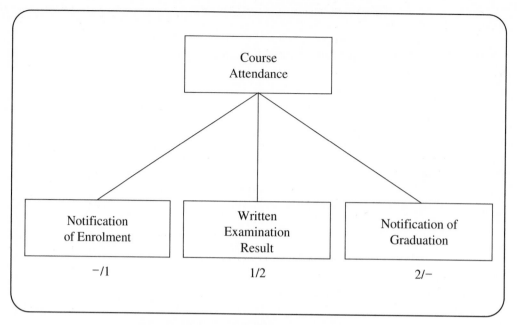

Figure 7.24 State indicators - sequence

A sequence of events will have the state indicator set to the value of the preceding effect (Figure 7.24).

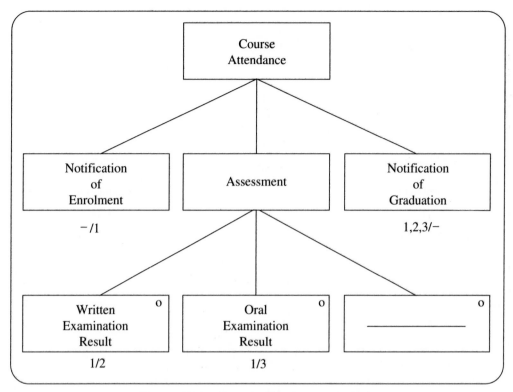

Figure 7.25 State indicators - selection

Alternatives within a selection structure will each have the same valid prior values. The event following the selection will have as valid prior values all the set to values within the selection structure. If a null event is included in the selection, the effects within the selection become optional and so the set to value of the effect preceding the selection will also be included in the valid prior values of the succeeding event (see Figure 7.25). Note that the null event does not need state indicators as it does not have an effect.

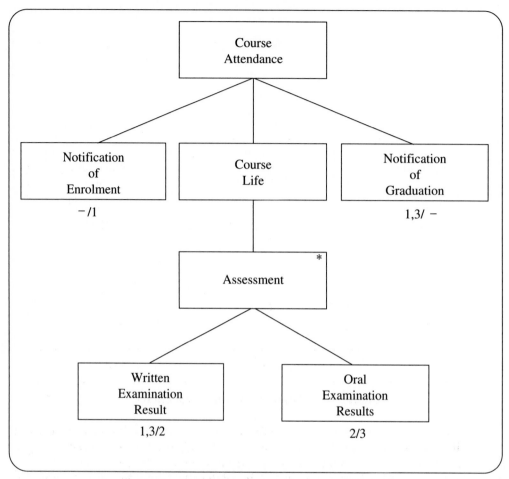

Figure 7.26   State indicators - iteration

Iterated events will include the set to state indicator value set by the last effect of the iteration, as in Figure 7.26. The event following the iterated effect must include the set to value of the effect preceding the iteration as iteration event(s) need not occur.

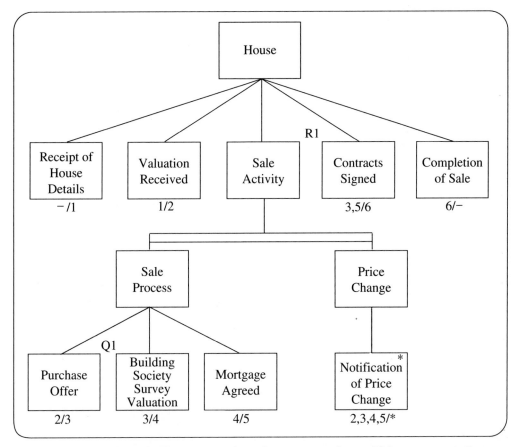

Figure 7.27 State indicators - quit and resume and parallel events notation

The Resume effect must have as a valid prior value the set to value of the Quitting effect, as will the effect diagramatically succeeding the quit (Figure 7.27).

Where parallel structures have been used, only one of the legs is considered as the main life of the entity. This is usually drawn as the first leg in the structure. It is only the effects on this main leg that may update the state indicator, known in this case as a primary state indicator. All the other legs of the parallel structure must leave the primary state indicator unchanged. This is indicated by use of an asterisk (*) as the set to value (Figure 7.27).

The valid prior values of the primary leg will follow normal conventions. On the subsidiary legs, the valid prior values will be all the set to values of the primary leg, plus valid set to values of any effect which can immediately precede the parallel structure.

If there is a need to control the events of any of the subsidiary legs, a subsidiary state indicator is introduced for each subsidiary leg. Each subsidiary state indicator must be treated as an additional attribute in the same manner as the primary state indicator and, in general, the same conventions apply in allocating values. A subsidiary state indicator is placed, in brackets, below the primary state indicator. The first effect within the secondary leg of the parallel structure has valid prior values associated with a null value and the last effect(s) in the leg. An example is shown in Figure 7.28.

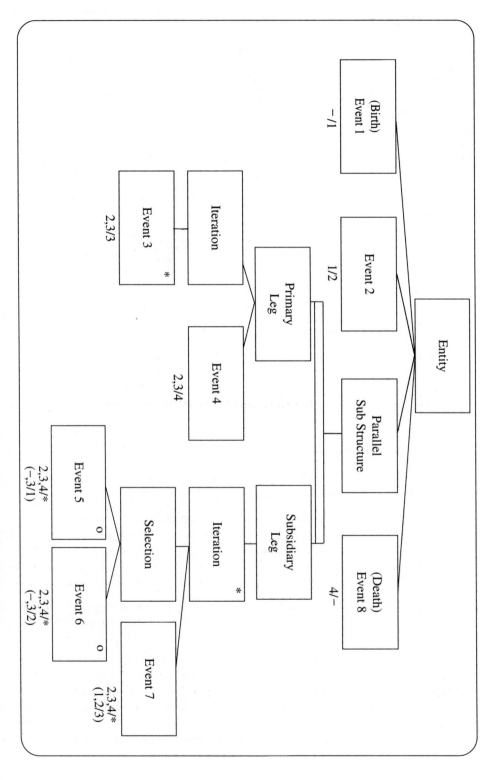

Figure 7.28   State indicators - monitoring a subsidiary leg of a parallel structure

7.7 COMPLETING THE CONSTRUCTION

7.7.1 Add Operations

The consideration of operations for each ELH is formalised by listing the operations at the bottom of the ELH or in a separate document and giving each a number. These can be checked against the Elementary Process Description.

Only major operations are considered at this stage; the detail is dealt with during the production of the update process models during the logical design activities. These major operations will include:

- store primary key value – birth effect;
- store initial attribute values – birth effect;
- store an attribute value – mid-life effect;
- store an attribute value using an expression – birth or mid-life effect;
- replace an attribute value (using expression) – mid-life effect;
- tie a detail entity occurrence to its master occurrence to make a relationship occurrence – birth or mid-life effect;
- cut a detail entity occurrence from its master occurrence to break a relationship occurrence – mid-life or death effect.

This list represents the minimum set of operations, although practitioners may feel the need to include additional types of operations. However, the intention is not to replicate the detail of other SSADM products such as the Elementary Process Description, but to aid the validation of the ELHs and the validation of the process model against the data model. To assist this validation, some practitioners might want to include a Gain detail for master entity operation and a Lose detail from master entity operation. Thus a check can be made that each Gain has a Tie and each Lose has a Cut operation and vice versa. The decision on whether to include the Gain and Lose operations may be dependent on the way that the proposed DBMS implements relationships.

The operations being considered are still logical, physical types of operations such as reading before modifying, accessing entities for navigation of the data model; error/exception handling, manipulating/sorting data, etc will be dealt with during production of the update process models.

Figures 7.29, 7.30, 7.31, 7.32 and 7.33 show the operations included on the ELHs for the entities Loan, Borrower, Book Copy, Reservation and Book. The Book ELH illustrates the problem associated with Gains and Lose operations. The End of year Analysis is where the Reservation details are lost, but there is no event to gain them. This would be the Notification of Reservation event but, since this does not have an effect on the stored Book data, there is no reason to include it in the Book ELH except for the sake of consistency and balance of the operations.

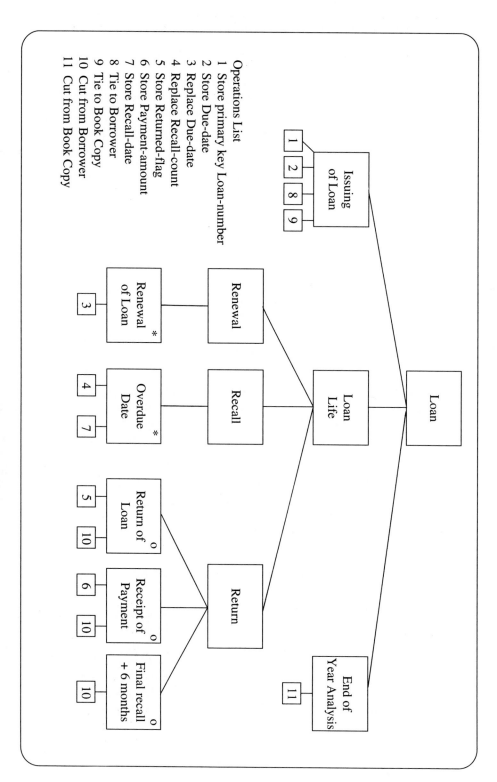

Figure 7.29   Operations - Loan ELH

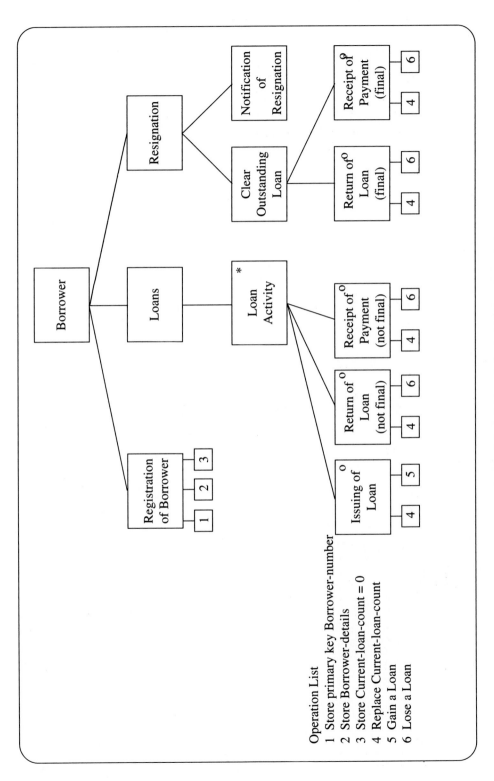

Figure 7.30 Operations - Borrower ELH

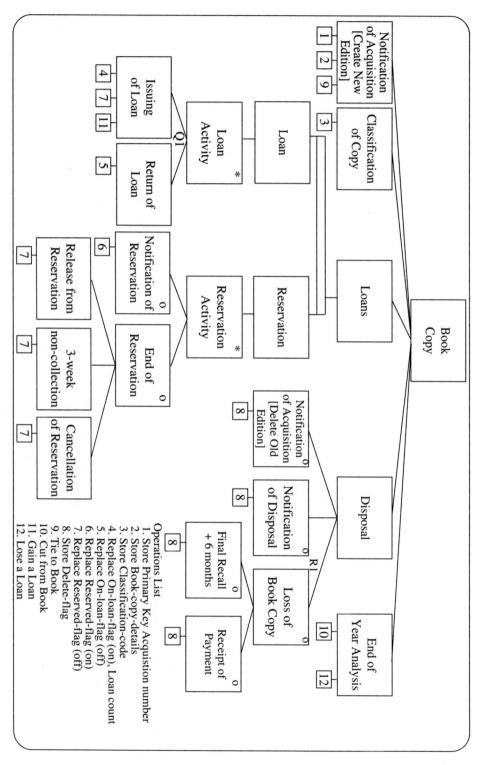

Figure 7.31 Operations - Book Copy ELH

Operations List
1. Store Primary Key Acquisition number
2. Store Book-copy-details
3. Store Classification-code
4. Replace On-loan-flag (on), Loan count
5. Replace On-loan-flag (off)
6. Replace Reserved-flag (on)
7. Replace Reserved-flag (off)
8. Store Delete-flag
9. Tie to Book
10. Cut from Book
11. Gain a Loan
12. Lose a Loan

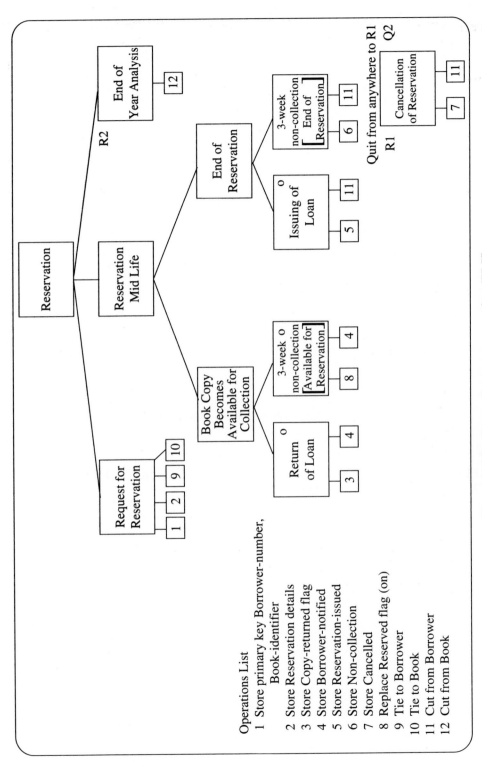

Figure 7.32 Operations - reservation ELH

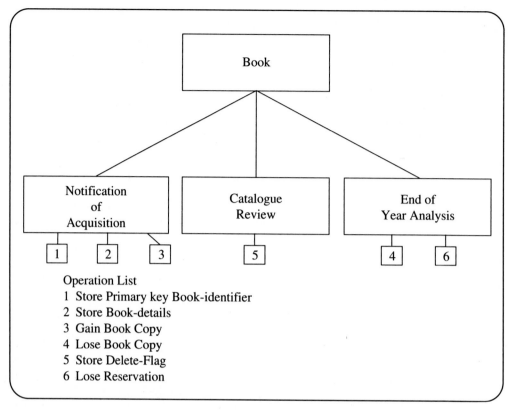

Figure 7.33   Operations - Book ELH

7.7.2   Add State Indicators

The ELHs developed in the previous section are shown in Figures 7.34, 7.35, 7.36, 7.37 and 7.38 with state indicators added.

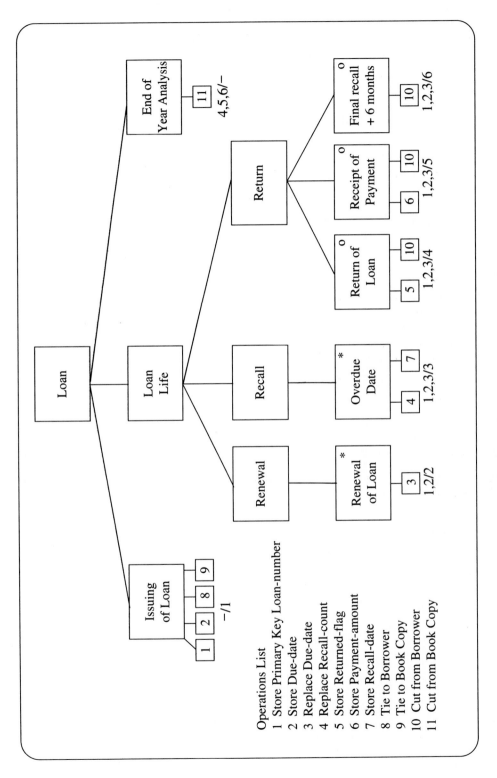

Figure 7.34 State indicators — Loan ELH

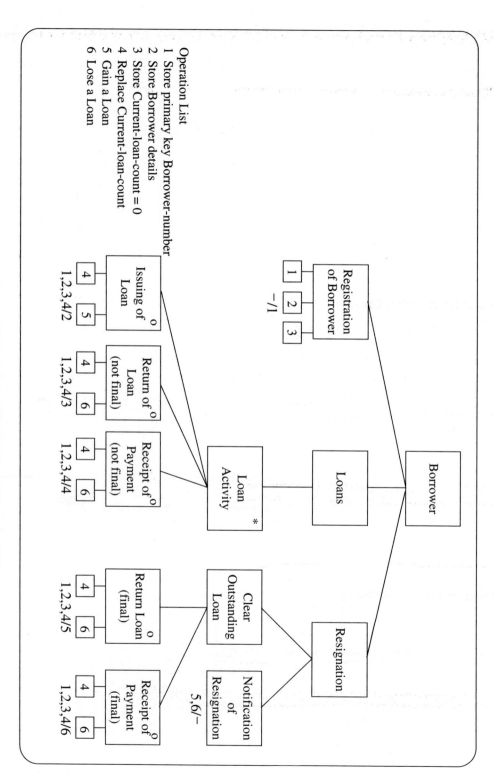

Figure 7.35   State indicators — Borrower ELH

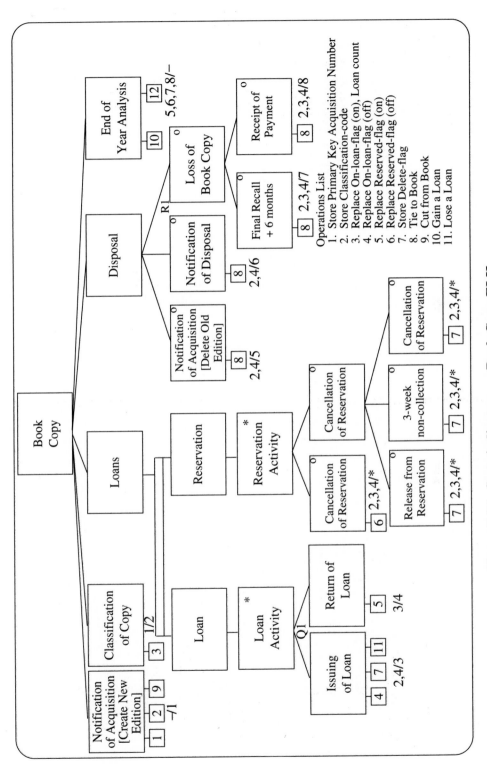

Figure 7.36   State indicators — Book Copy ELH

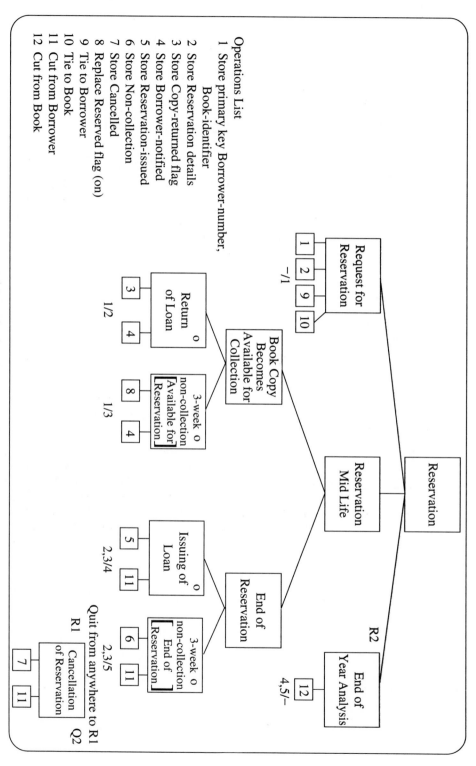

Figure 7.37   State indicators — Reservation ELH

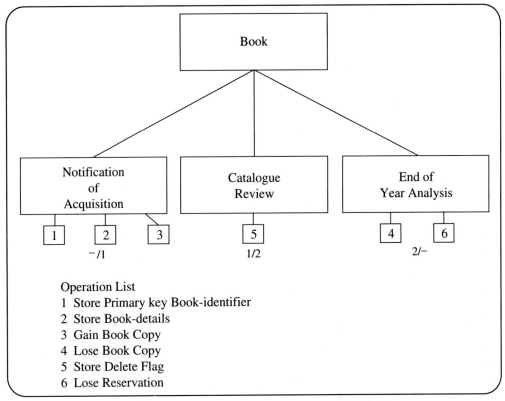

Figure 7.38 State indicators — Book ELH

The inclusion of state indicators on the Book Copy ELH highlights a sequencing problem not previously identified. The two events associated with the loss of a book may only abnormally follow the Issuing of Loan event. However, the state indicators show that their position in the diagram allows them to follow the Classification of Copy and the Return of Loan events as an allowable sequence. This is obviously not possible. A resolution to this problem is to place the loss of book copy as a separate substructure. Figure 7.39 shows the corrected ELH for Book Copy.

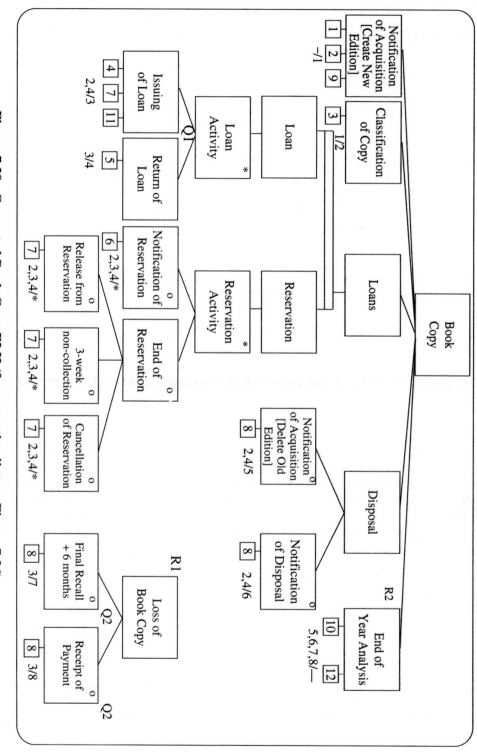

Figure 7.39   Corrected Book Copy ELH (for operation list see Figure 7.36)

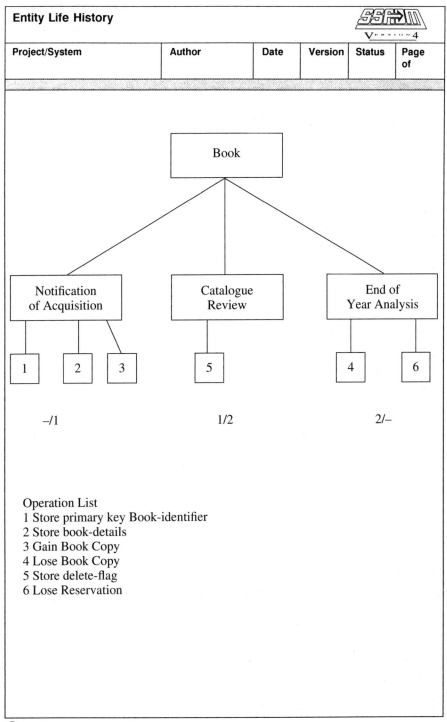

Figure 7.40   Entity Life History

7.8   DOCUMENTATION

Documentation of ELHs will include:

- – a set of ELH diagrams (Figure 7.40);
- – a list of operations for each ELH (Figure 7.40);
- – a description of the events (on the Function Description, see Chapter 5).

7.9   RELATIONSHIP WITH OTHER MODELS

The components of the Logical Data Model, particularly the LDS and the Entity Descriptions, are important inputs to the construction of Entity Life Histories. After all, it is the LDM which describes the entities on which the ELHs are based. The detailed analysis of the ELH technique develops further understanding of the data and this will feed back into the LDM. This results in adding new entities or splitting existing entities, removing entities as a result of merging, changing relationships or adding relationships, adding or removing attributes. It is also ELH analysis that adds state indicators and their meanings to the Entity Descriptions.

Events are identified from the Data Flow Model and recorded in the Function Descriptions. The Data Store/Entity Cross Reference is used to help identify which entities are affected by each event when producing the Event/Entity Matrix.

The knowledge gained in constructing the Entity Life Histories is likely to provide significant feedback into the Function Definition. This will usually result in the creation of new functions and the amendment of existing ones in the light of the greater understanding gained from constructing the ELHs. Indeed, it is difficult to be confident about function definition until after the construction of the ELHs.

The events identified during the ELH analysis are the major input to the Effect Correspondence Diagrams, the second phase of Entity-Event Modelling. The ECD documents the effect of each event on the system data (see Chapter 8).

ELHs, showing the sequence of required processing and identified operations, are also major contributors to the Logical Database Process design (see Chapter 10).

7.10   SUMMARY

This first stage of Entity-Event Modelling, Entity Life History analysis, is concerned with:

- – identifying events that cause changes in stored data;
- – establishing the sequence of these events;
- – identifying major operations associated with each effect;
- – defining valid states for entity occurrences.

Effects are documented on ELHs and given the name of the event causing the effect. Operations relating to each event are also modelled on the ELH.

Entity Life History analysis is concerned with eliciting and modelling the sequences and constraints of the business system. The users involved in entity-event modelling must know the business rules, be able to describe unusual business situations, define how to handle error situations and be aware of the way different business areas interact and the movement of data between them.

Entity Life Histories have developed only a little from Version 3 to Version 4. Version 3 recognised that it was necessary to qualify events when the effect could vary according to circumstances or to enforce a business rule. Version 4 introduces the additional concepts of mutually exclusive events, entity roles and operations. The former concepts are required for Effect Correspondence Diagramming, the latter is a powerful aid to validating the LDM and ensuring that the business rules of the required system are properly represented in the ELH.

It is also worth noting that the use of the Quit and Resume notation has been more specifically defined in Version 4 and care should be taken by anyone migrating from Version 3 to Version 4 to ensure their use of this construct is consistent.

However, perhaps the most significant aspect of Entity Life Histories is their promotion as a more central model in SSADM system development. Many Version 3 practitioners perceived them as an optional model which did not provide any further basis for design. This is not the case in Version 4, an issue we return to in the final chapter.

8 Enquiry Access Paths and Effect Correspondence Diagrams

8.1 INTRODUCTION

Each update and enquiry function requires an access path through the Logical Data Structure (LDS) to locate the necessary data. In SSADM, the enquiries are modelled with an Enquiry Access Path and the update functions are defined on Effect Correspondence Diagrams.

8.1.1 Enquiry Access Paths

An Enquiry Access Path shows the route through the LDS required to obtain the data needed to fulfil an enquiry or to fulfil the enquiry component of an update function. Each Enquiry Access Path is described diagrammatically with one diagram for each enquiry.

The Enquiry Access Paths are initially developed to validate that the LDM can support the enquiry requirements. They are subsequently used in the development of the Enquiry Process Models.

An Enquiry Access Path may also be defined for enquiries within update functions. These are enquiries made before or after all the processing of an event takes place.

8.1.2 Effect Correspondence Diagrams

The Effect Correspondence Diagram (ECD) is the second modelling technique used in Entity-Event Modelling. An ECD is drawn for each event. It demonstrates update access paths by showing the way in which effects on entities are related to each other. ECDs are derived from the set of Entity Life Histories. They are subsequently used in the development of the Update Process Models.

Effect Correspondence Diagrams are similar to Enquiry Access Paths but do not contain all the explicit navigation detail.

8.2 MODELLING NOTATION – ENQUIRY ACCESS PATHS

The Enquiry Access Path is drawn using the basic notation described in Chapter 2, Structure Diagrams. Each entity is shown on the Enquiry Access Path using the 'soft' box notation.

8.2.1 Iteration Structure

The iteration structure is required when more than one entity occurrence needs to be read (accessed). For example, for the enquiry 'List all customers' the Customer entity is placed in the iterated box and the box above is labelled 'Set of' with the entity name. This convention, used throughout Enquiry Access Path production, is shown in Figure 8.1.

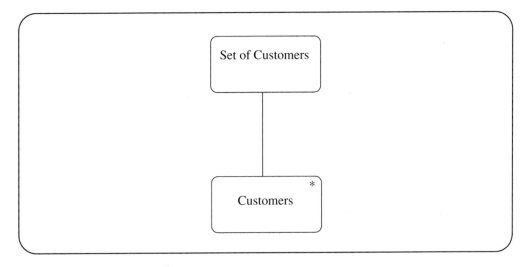

Figure 8.1   Enquiry access path–iteration structure

8.2.2   Selection Structure

The selection structure is used when the access path involves exclusive relationships. For example, Figure 8.2 is a view of an LDS showing the entities which hold the data needed to answer the enquiry 'What are the sales terms for each customer?'. Figure 8.3 shows the selection structure derived from this LDS to answer that query. For each Customer, access is made either to their Credit Terms or Cash Terms as appropriate.

Access correspondence (single headed) arrows are used to link enquiry access structures to show the order of access to the entities.

A single headed arrow is also used to indicate the first entity in the access sequence, as in Figure 8.4. This diagram illustrates the Enquiry Access Path for the enquiry 'List all the customers for a given sales area with details of their credit terms or cash terms as appropriate'.

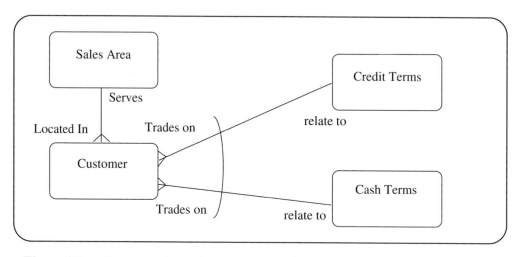

Figure 8.2   View of an LDS for the enquiry 'What are the sales terms for each customer?'

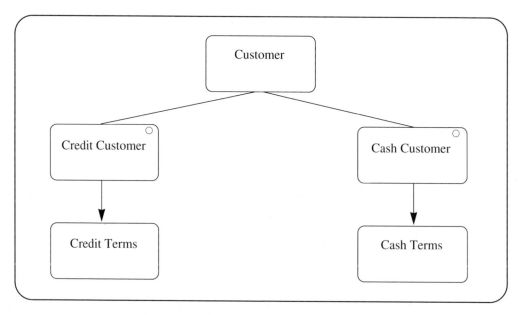

Figure 8.3   Enquiry access path – selection structure

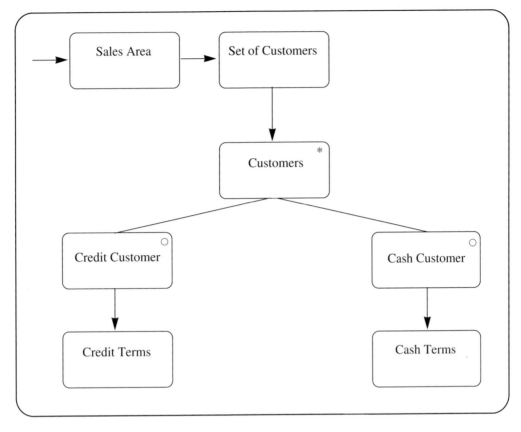

Figure 8.4   Enquiry access path for the enquiry 'List all the customers for a given sales area with details of their credit or cash terms as appropriate'

8.3   CONSTRUCTION OF AN ENQUIRY ACCESS PATH MODEL

The basic steps in developing an Enquiry Access Path for each named enquiry are as follows:

- specify the enquiry trigger;
- identify the entities to be accessed;
- draw the required view of the LDM;
- redraw the required view as an Enquiry Access Path;
- identify the entry point entity;
- validate the diagram;
- document all entry points.

Each enquiry must be identified by a unique name and it is this name which is used as the name of the access path.

Using the example of the Public Library system (the LDS is reproduced as Figure 8.5) a possible enquiry is to 'List for each junior library member the books they have on loan and the date the loan is due to be returned'. The enquiry is identified by 'List loans by junior borrowers'.

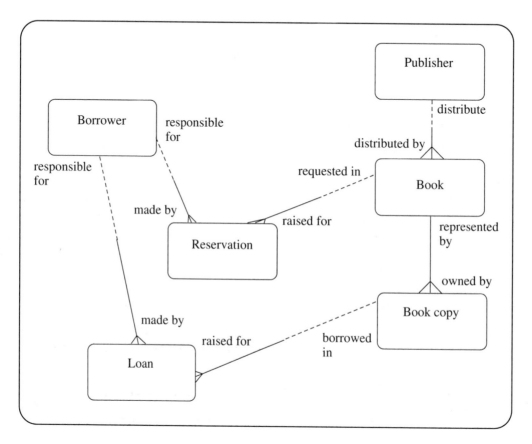

Figure 8.5   LDS for a public library system

8.3.1   Specify the enquiry trigger

The enquiry trigger comprises the data items entered into the enquiry and this is normally the key of the entity which is the entry point to the LDM. In addition there may be some selection criteria governing the retrieval of detail entities.

To list loans by junior borrowers the only input data item will be the Membership-type attribute. There will be no additional selection criteria.

8.3.2   Identify the Entities to be Accessed

The entities that have to be accessed to provide the data for the enquiry output can be informally derived from the I/O Structure for the enquiry.

To list loans for junior members these entities are:

- Borrower;
- Loan;
- Book;
- Book Copy.

8.3.3   Draw the required view of the LDM

The view of the LDM is the subset of the LDS which includes:

- the entities identified in the previous step;
- the relationships connecting these entities.

The relationships represent the access paths between the entities for the enquiry. When drawing the view of the LDS ensure that each relationship (access):

- from a master entity to a detail entity is drawn vertically;
- from detail to master entity is drawn horizontally.

Figure 8.6 shows the required view for the enquiry.

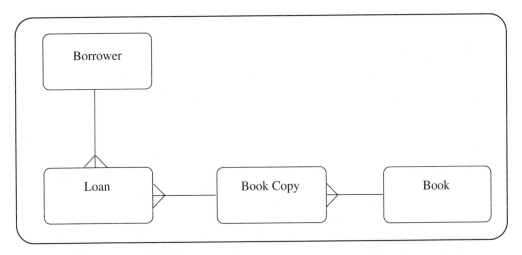

Figure 8.6   Required view of LDS for 'List loans by junior borrowers'

8.3.4   Redraw the Required View as an Enquiry Access Path

The required LDS view is redrawn using the Enquiry Access Path notation (see Figure 8.7). Entry is via membership type into Borrower (see next step). For each junior member the set of all their Loans is accessed for a Due-date of return. This is indicated by the iteration structure. For each of these loans the master entity Book Copy is accessed in order to gain access to the Book entity where the book details are stored.

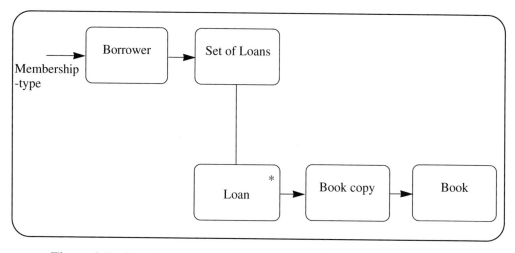

Figure 8.7   Enquiry access path for 'List loans by junior borrowers'

8.3.5   Identify the Entry Point Entity

There are three kinds of entry point:

- via a primary key (identifier);
- via a non-key attribute of an entity or attribute that forms part of the primary key;
- to all occurrences of an entity.

In the first two cases, the attribute name is placed against the entry point arrow. In the third case, the arrow is unnamed indicating that there is no selection. Figure 8.7 shows the entry point of Membership-type, a non-key attribute, into Borrower.

It may be felt at this stage that Membership-type is a sufficiently significant attribute to merit an additional entity Membership. Its identifier Membership-type would then become a foreign key in Borrower. In these circumstances the LDS view would be as in Figure 8.8 and the Enquiry Access Path as in Figure 8.9. This illustrates how Enquiry Access Path modelling can feed back into the Logical Data Model.

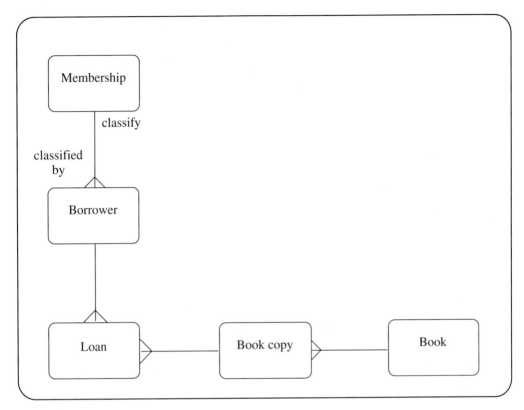

Figure 8.8   Updated LDS view for 'List loans by junior borrowers'

A foreign key of an entity should not be used as an initial entry point. If it is necessary to access a detail entity by one of its foreign keys, then the entry point should be the primary key of its master entity. In effect, this enforces an existence check on the value of the foreign key.

Moreover, an entry point via a combination of foreign keys should be represented as an entry point into the relevant master with this combination as its primary key. If such an entity does not exist, then consideration should be given to including one in the LDS. For example, the LDS in Figure 8.10 may be a view to support the enquiry 'List the patients in a consultant's care and, where appropriate, the ward to which they have been admitted'. Using Consultant as the entry point the result of the enquiry would not be grouped by patient within ward for each Consultant. However, by including the Consultant Ward entity (Figure 8.11), and using it as the entry point a path is provided into the Patient entity on a combined foreign key and this enables the required grouping.

Further discussion with the user may reveal that this new entity is also justified in business terms. It can be used to hold such details as the date when a consultant was first allocated beds in the ward, the number of beds allocated to the consultant and the normal time of ward rounds.

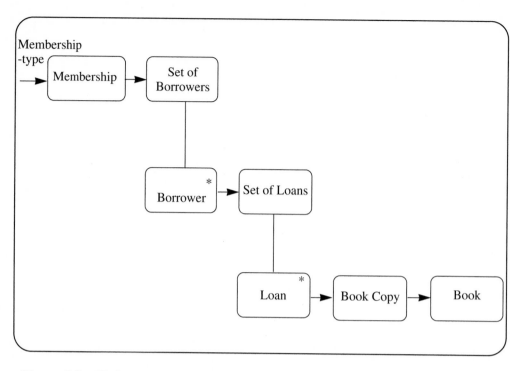

Figure 8.9   Updated enquiry access path for 'List loans by junior borrowers'

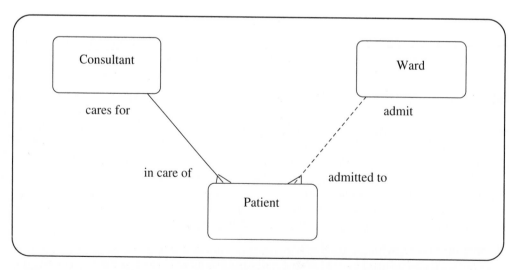

Figure 8.10   LDS view to list of patients in care of a consultant and the ward, where appropriate, to which they have been admitted

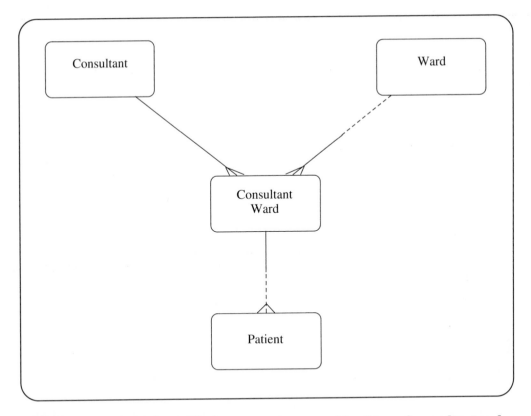

Figure 8.11   Amended LDS view to access on combined key of consultant and ward

Alternatively, if the Consultant Ward entity is not significant in business terms and carries no attributes in its own right, a process oriented solution to the enquiry, such as sort the output data, might be preferred. In this instance the Enquiry Access Path should be annotated with the sort requirement.

8.3.6   Validate the Diagram

The Enquiry Access Paths needs to be checked for:

- use of read operations;
- access rights;
- connection traps.

After the entry points are identified, a check is required on the access path to ensure that all the required data can be accessed using only the operations:

- read entity directly using identifier (key attribute);

- read next detail of a master entity;

- read master entity of a detail entity.

If this is not possible then it may be necessary to introduce a new entity, an alternative entry point, or to modify or add relationships on the LDS to match the enquiry requirements. Alternatively, it may be sufficient to document some process-oriented solution, such as a sort.

It is also important to ensure that the access rights of the users designated to make the enquiry are consistent with those defined in the Entity and Attribute Descriptions.

Even if the access path is correctly structured, it cannot be assumed that it will be possible to complete the enquiry path or that the path will give a correct answer to the enquiry. These problems usually arise because of a misinterpretation or misuse of relationships on the LDS. These potential misinterpretations are known as connection traps.

There are three types of connection traps:

- chasm trap;

- relationship fan trap;

- misinterpretation trap.

The existence of a relationship between entities is not a guarantee of a path actually existing from all entity occurrences. For example, in Figure 8.11 a patient is linked to consultant via the Consultant Ward entity, so it would appear that the consultant for each patient can be identified. However, the optionality of the relationship means that there are patients which have not been admitted to a ward. Thus it is not possible to identify their consultant. For these patients (out-patients) it is necessary to have an additional relationship to link them to their consultant, as in Figure 8.12. This type of connection trap is called a chasm trap.

A relationship fan exists when two or more relationship occurrences fan out from the same entity occurrence. For example, in Figure 8.5 a Borrower is responsible for a Reservation which is raised for a Book and each Book may have many Reservations requested for it. There are many Book Copies for each Book and each Book Copy is flagged as reserved until one is available. It is then held awaiting collection from the Borrower. However, the occurrence diagram (Figure 8.13) shows that there is no way of determining with which Reservation this flagged Book Copy is associated.

The information solution to this fan trap is to have an additional relationship between Reservation and Book Copy to link the entity occurrences. The Book Copy available for Reservation relationship, illustrated in Figure 8.14, means that the reservation holding the book copy, and hence the borrower, can now be modelled on an Enquiry Access Path and unnecessary processing is avoided. It also means that the processing to identify the next reservation for a returned book copy is also simplified.

Consideration also needs to be given as to how this one-to-one relationship is to be represented.

The entity identifiers are not the same and so it is not appropriate to merge the entities.

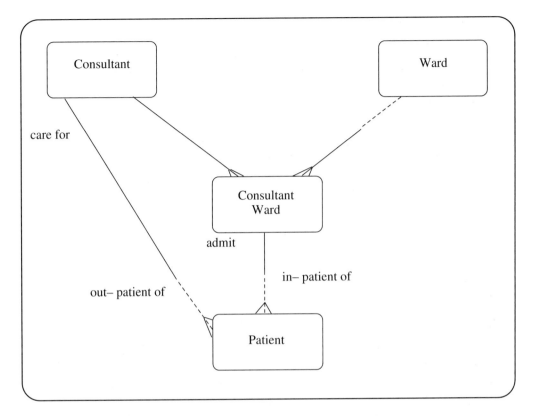

Figure 8.12   Solution to a chasm trap

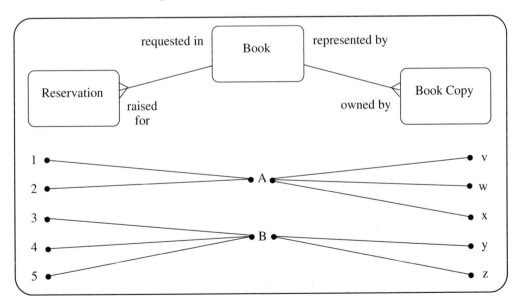

Figure 8.13   Occurrence diagram showing a relationship fan trap

Book Copy is more permanent in the structure and its occurrences are generally included in the data structure earlier in time than the reservation occurrences associated with them. Thus it seems sensible to designate Book Copy the master entity and place its identifier, Acquisition-number, as a foreign key in the reservation entity.

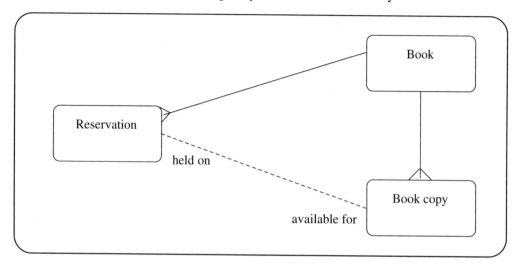

Figure 8.14   Solving the relationship fan trap using an additional relationship

The business context also requires consideration. If the number of reservations compared to the number of books is small, or if only a few of the many books are subject to reservation, then this solution is appropriate. However, if the library has a fairly small book stock and the number of reservations for each book is high, then the preferred solution may be to make Reservation the master and Book Copy the detail. To implement this solution, it might also be desirable to create a Reservation-number to uniquely identify each reservation rather than the current composite identifier of Borrower-number, Book-identifier and Date-of-reservation. Each of the possible solutions has an effect on storage volumes and process and enquiry efficiency.

A Relationship fan does not always cause a problem; it depends on the business context of the process or enquiry. For example, when a reservation for a book is raised the path from Reservation to Book to Book Copy also traverses a relationship fan. However, in this case, all the Book Copies of the Book do need to be identified in order to flag them all for reservation.

Care must be taken that the definition of the relationships are understood and are relevant to the enquiry, otherwise the data output from the enquiry may actually be the answer to a different enquiry. Conversely, care must also be taken that the exact nature of the enquiry which the Business user defines is clearly understood. For example, the Enquiry Access Path for 'List loans for junior borrowers' (Figure 8.9) has two interpretations. The first is to list all loans since the last end of year analysis. The second is to list only current loans, that is loans in which the book copy has not yet been returned. The relationship name does not help decide which enquiry will be answered. The problem should be resolved by the description given on the Relationship Description Form (Figure 4.21) where it is clearly stated that all loans that a borrower has been, and is currently responsible for, are included in the relationship.

If the enquiry requires only current loans, then a modification is required to the Enquiry Access Path. Figure 8.15 shows that the set of loans for a Borrower can be considered as two subsets, the returned (historic) loans and the current loans. It is only the current loans for which the Book details are required. Confirmation of which enquiry was intended must be sought from the user.

The evolution of the LDS to resolve various business requirements and syntactical problems of Enquiry Access Paths has illustrated another important point. If any changes are made to the LDM while developing an access path, it is necessary to revalidate all other Enquiry Access Paths to ensure that they are still giving valid solutions to the defined enquiries.

8.3.7   Document All Entry Points

All the entry points identified during the development of the Enquiry Access Paths are consolidated onto a single copy of the LDS and annotated with the relevant attributes. Each entity may have several entry points associated with it. This information will be used in the physical data design.

8.4   MODELLING NOTATION – EFFECT CORRESPONDENCE DIAGRAMS

It is useful to begin this section by reviewing a number of concepts introduced in the previous chapter:

- event – causes an effect on one or more entities;
- effect – the change caused by an event on a single entity;
- mutually exclusive effect – an event affects one entity occurrence in one of several mutually exclusive ways;
- entity role – a single event affects more than one occurrence of an entity and the effects are different for each entity occurrence.

8.4.1   Notation

Each Effect Correspondence Diagram has the name of an event as the heading. Effects are represented by round-cornered boxes and the types of structures used are similar to the Structure Diagrams, although they are combined in a different way.

8.4.2   Single Effect on Single Entity Occurrence

If an event has a single effect on just one occurrence of an entity then the effect box contains the name of the entity affected by the event. For example, if the event Transaction

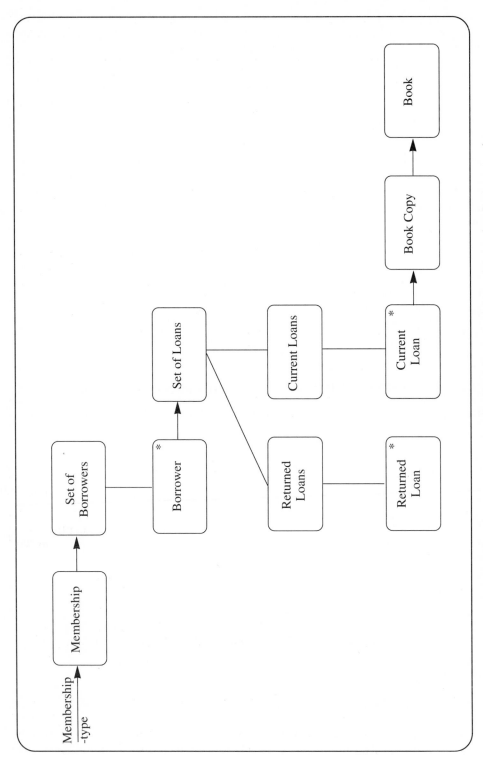

Figure 8.15   Enquiry access path for 'List *current* loans by junior borrowers'

has an effect on the Statement Line entity the effect, named Statement Line, occurs on the ECD as shown in Figure 8.16.

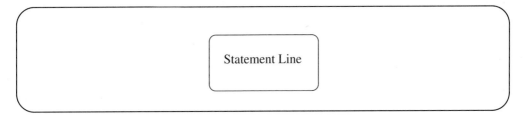

Figure 8.16   ECD – single effect

8.4.3   Selection

The selection notation is used when an event has an effect on a single entity occurrence in two or more mutually exclusive ways. For example, consider the event Transaction; its effect on the entity Account is found to be different depending on whether the transaction is a payment or a withdrawal. Figure 8.17 shows the mutually exclusive effects on the entity Account labelled with the qualified event names. Whenever the Transaction event occurs, the effect on the Account entity will be the operations associated with either a payment or a withdrawal but not both.

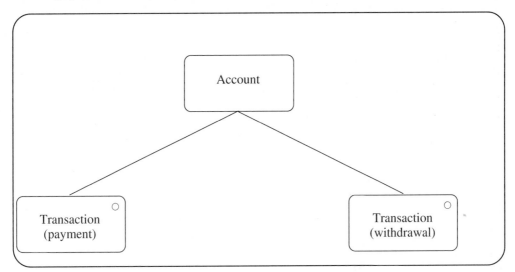

Figure 8.17   ECD – selection of effects

8.4.4   Iteration

If a single instance of an event causes more than one occurrence of a particular entity to be changed then the iteration of the event on the set of entity occurrences is depicted using the structure in Figure 8.18. The asterisk indicates that the effect on the entity Fund is iterated and a Set of Funds, that is a set of entity occurrences of the type Fund, are changed when the event Transaction occurs. Money associated with the transactions may be paid into or withdrawn from one or more different Funds.

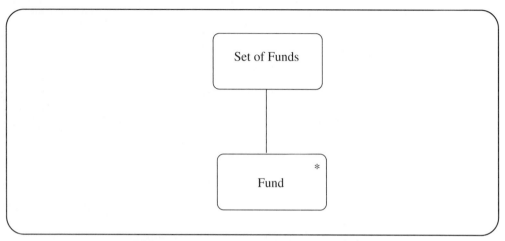

Figure 8.18   ECD – iteration of effects

8.4.5   Simultaneous Effects

There are occasions when a single instance of an event affects more than one occurrence of an entity, each occurrence being affected in a different way. For Example, if the business rules state that a Patient can only be registered with one Doctor at any one time, then the effect of the event Notification of Registration should have the effect of deleting the old Registration and creating a new Registration. Thus the events are logically simultaneous. Both effects can be identified from the ELH and must be shown on the ECD as illustrated in Figure 8.19, where the entity name has been qualified with its various roles.

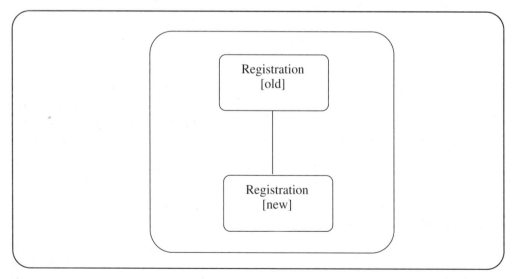

Figure 8.19   ECD – simultaneous effects

8.4.6   One to One Correspondence

The one to one correspondence between the effects of a single event is shown by a bi-directional arrow. There will be a one to one correspondence when an instance of an event causes one occurrence of an effect on one entity occurrence and also one occurrence of another effect on another entity occurrence. This brings together all the effects associated with one event into a single diagram.

Figure 8.20 shows that for one occurrence of the Transaction event, one or other of the mutually exclusive effects will take place on the Account entity. In addition, there is a corresponding effect on a Statement Line occurrence and a corresponding effect on a Set of Fund entity occurrences.

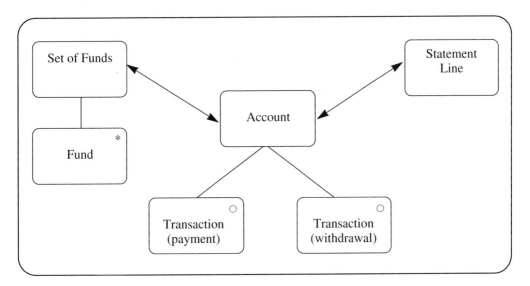

Figure 8.20   ECD for the Transaction event

8.5   CONSTRUCTION OF EFFECT CORRESPONDENCE DIAGRAMS

Effect Correspondence Diagramming is an important technique for validating the ELHs. ECDs can be started during ELH analysis and developed in parallel with the ELH diagrams. An Effect Correspondence Diagram is drawn for each event identified on the Entity Life Histories.

For each event the steps in deriving an Effect Correspondence Diagram are:

- draw a box representing each entity affected by the event;

- draw separate boxes for simultaneous effects;

- include mutually exclusive effects;

- add iterations;

- add one to one correspondences between effects;

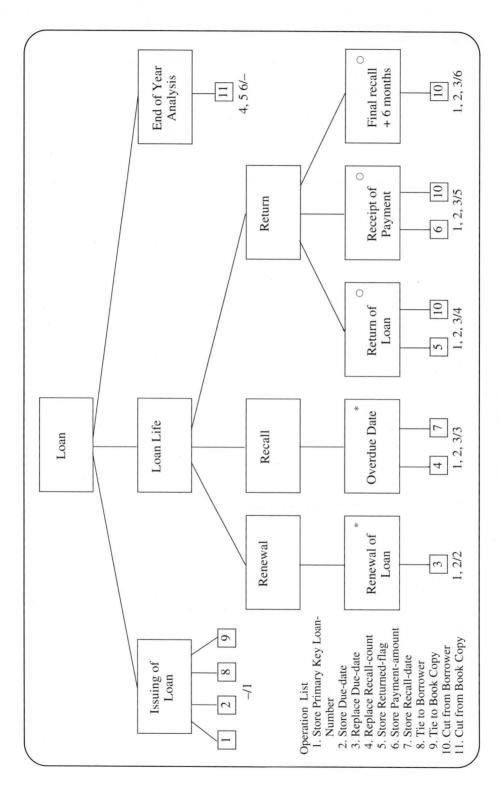

Figure 8.21   Loan ELH

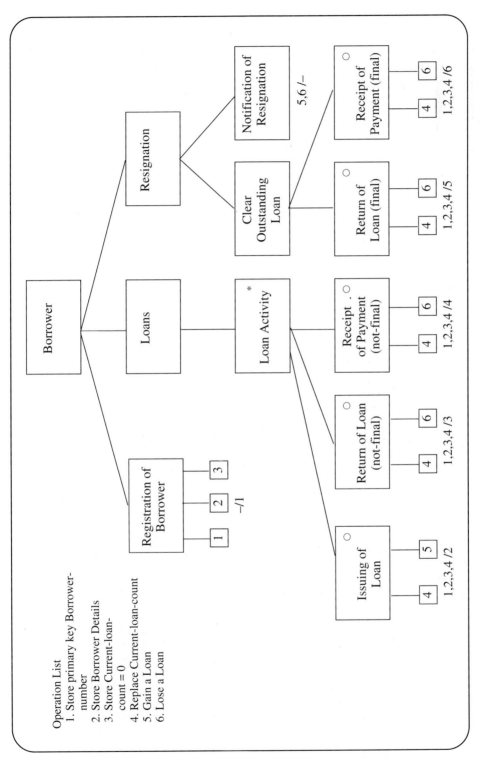

Operation List
1. Store primary key Borrower-number
2. Store Borrower Details
3. Store Current-loan-count = 0
4. Replace Current-loan-count
5. Gain a Loan
6. Lose a Loan

Figure 8.22 Borrower ELH

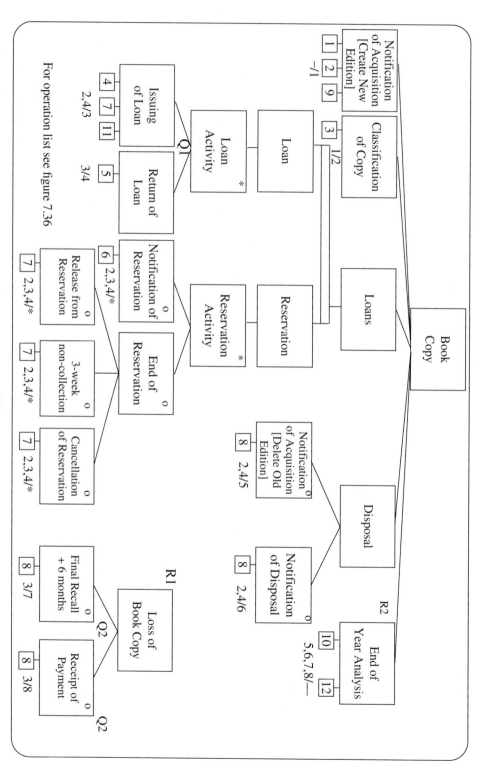

Figure 8.23　Book Copy ELH

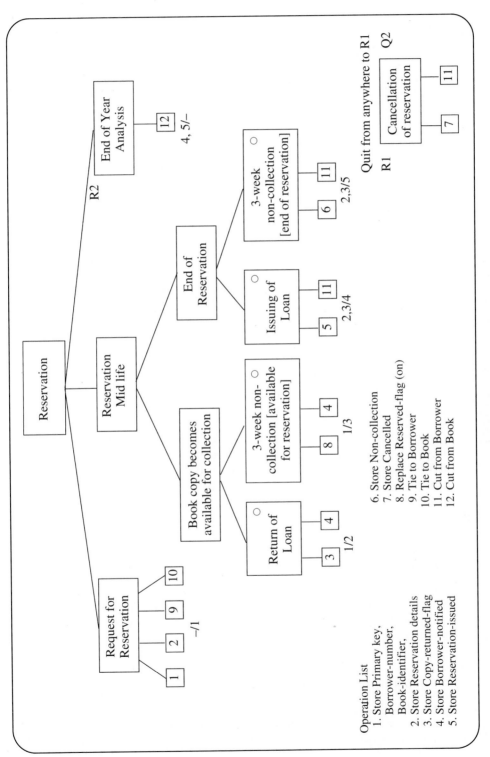

Figure 8.24 Reservation ELH

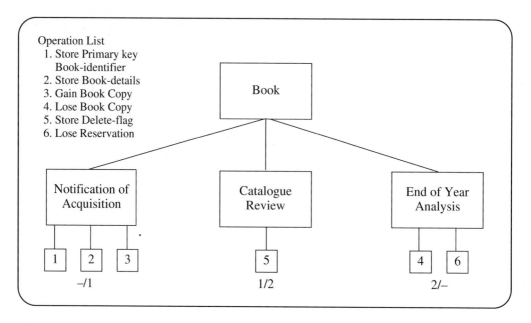

Figure 8.25   Book ELH

- merge iterative effects;

- add non-updated entities;

- add event data.

ECDs for the events Return of Loan, Notification of Reservation, 3-Week Non-collection and Notification of Acquisition will be developed to illustrate some of the principles of ECD construction. The ELHs for Loan, Borrower, Book Copy, Reservation and Book are duplicated from Chapter 7 in Figures 8.21, 8.22, 8.23, 8.24 and 8.25 for reference. The ELHs have not been updated as a result of validation during the development of the Enquiry Access Paths.

8.5.1   Draw a Box Representing Each Entity Affected by the Event

An ECD form is used for each event (see Figure 8.41). The affected entities are identified by looking to see which ELH includes the specified event. The entities affected by the named event are then drawn using 'soft' boxes on the form. It is useful to place the entities in a similar layout to their position on the LDS.

8.5.2   Draw Separate Boxes for Simultaneous Effects

A simultaneous effect occurs when an instance of an event may logically affect more than one entity occurrence, each occurrence being affected in a different way. On the ELH this will be shown as the same event name qualified with different roles. The ECD for 3-Week Non-collection is shown in Figure 8.26 and illustrates simultaneous effects on the Reservation entity.

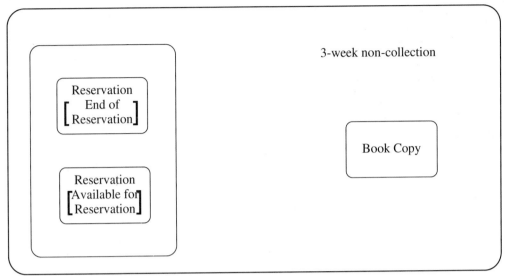

Figure 8.26   ECD showing simultaneous effects

8.5.3   Include Mutually Exclusive Effects

An event which has two or more mutually exclusive effects on an entity is shown as a selection on the ECD. Mutually exclusive events can be recognised on an ELH as dupli-cated events qualified by exclusivity. Figure 8.27 shows the effect of Return of Loan on the Borrower entity depending on whether it is a 'not final' occurrence or a 'final' instance of the event.

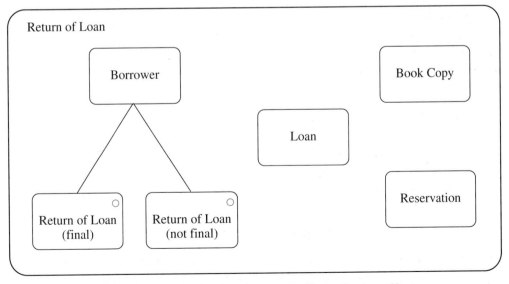

Figure 8.27   ECD showing mutually exclusive effects

8.5.4   Add Iterations

Iteration occurs where an event causes a set of entity occurrences to be updated by the same effect. Figure 8.28 shows the Notification of Reservation ECD in which a reservation of a book causes a set of Book Copy to be affected. The iterative effect is deduced from the business rules and is confirmed by observing that Book Copy is the detail of Book on the LDS.

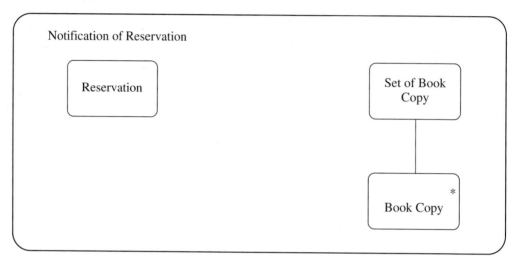

Figure 8.28   ECD showing iteration of effects

There are also iterative effects caused by the Notification of Acquisition event. When a new edition of a book is acquired, all the set of Book Copies of the previous editions are disposed of and the set of the new edition of Book Copy occurrences are created. This is illustrated in Figure 8.29.

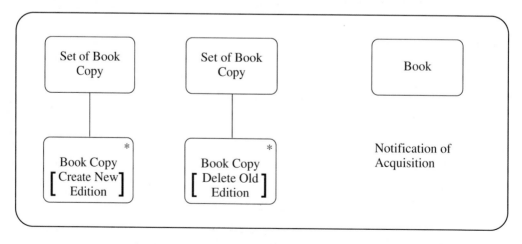

Figure 8.29   ECD showing iteration of effects

8.5.5   Add One to One Correspondences between Effects

Bi-directional arrows are added to the diagram where an effect on an occurrence of one entity type corresponds to an effect on an occurrence or occurrences of another entity type. One to one correspondences occur when one entity is a detail of another, indicating access from detail to master, as in Figure 8.30. Accessing from master to detail the one to one correspondence may be with a set of the entity as in Figure 8.31, Notification of Reservation. Here the access from a single Reservation is via a Book and Book is master of Book Copy. (The Book entity is not included in the diagram at this stage because it is not updated and no data is retrieved from it.) It may also be possible that access from master to detail will require the update of a single occurrence of the detail. In Figure 8.32, Return of Loan, access of the single Reservation occurrence, which may now be fulfilled, is required via Book Copy.

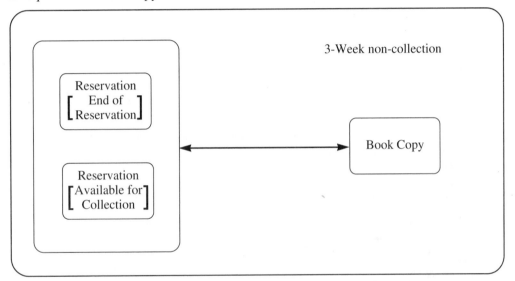

Figure 8.30   One-to-one correspondence – 3-week non-collection ECD

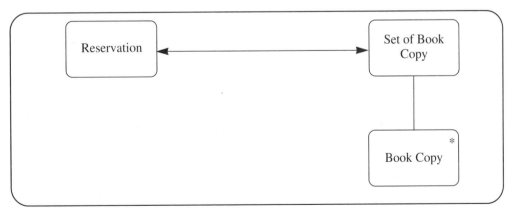

Figure 8.31   One-to-one correspondence – notification of reservation ECD

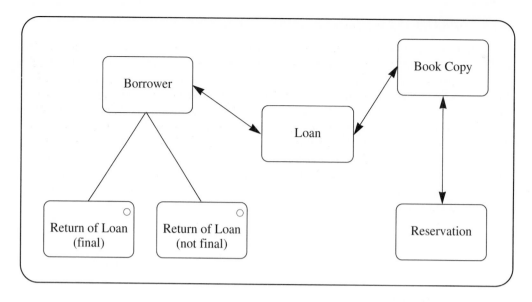

Figure 8.32   One-to-one correspondences - return of loan ECD

Figure 8.33 illustrates a further example of a one to one correspondence by extending the example with an additional event Withdrawal. When the Withdrawal event occurs, a Book occurrence is flagged as withdrawn and copies of the book are temporarily withdrawn from the loan stock. This means that for each Book Copy the current Loan, if any, has to be identified and recalled.

Thus there is a one to one correspondence of each of the identified Book Copies with just one of its Loan entity occurrences.

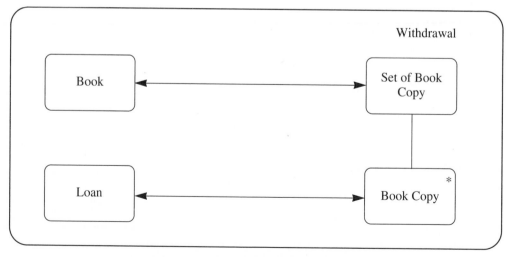

Figure 8.33   Possible withdrawal ECD

8.5.6   Merge Iterative Effects

If an entity is affected by an event in more than one iterative way and access to the entity is via the same relationship, then the effects can be merged into a single structure. For example, Notification of Acquisition, Figure 8.29, has more than one iteration of Book Copy. This is identified on the ELH for Book Copy as a type of entity role because the event has two different effects. Each different effect in this instance is on a set of entity occurences. Acquisition of a new edition creates a set of new occurrences of Book Copy, while all the occurrences of the old editions are disposed of (deleted). There may not be the same number of new editions as there were of the old editions. Merging the iterations gives the two possible structures illustrated in Figure 8.34.

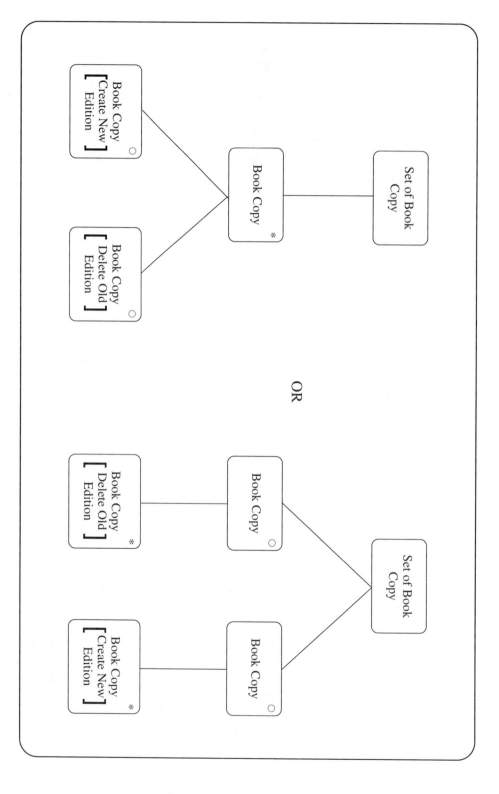

Figure 8.34　Merging iterative effects

The first represents a set of iterative changes that can include Notification of Acquisition [create new edition] and Notification of Acquisition [delete old edition]. The second represents an instance of an event causing either a set of changes representing Notification of Acquisition [create new edition] or a set of changes representing Notification of Acquisition [delete old edition]. It is the first structure which is required in the Public Library System and the complete ECD is shown in Figure 8.35. However, if the business rules been different the alternative merging of iterative effects could have been a valid structure.

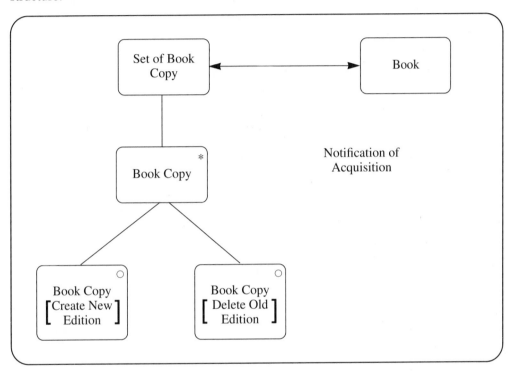

Figure 8.35   ECD – notification of acquisition

8.5.7   Add Non-updated Entities

To complete the ECDs, it is now necessary to include nodes to represent entities which need to be accessed to traverse the data structure, and entities from which data is retrieved but which are not updated.

To establish the access route through the LDM for each event, it may be helpful to produce a type of Enquiry Access Path. If it is found that additional entities are required to facilitate access, then these must be included on the diagram in the appropriate place. Using the unvalidated LDS in Figure 8.5, it can be seen that the ECDs developed for Notification of Reservation, Return of Loan and 3-week Non-collection all need Book including in their structure to enable navigation between Book Copy and Reservation.

It is also necessary to check that all the data required to produce the output defined in the I/O Structure for this event is included as attributes in the Entity Descriptions of the entities in the diagram. If additional entities have to be accessed but are not changed in any way, then these must also be added to the diagram and access validated.

8.5.8   Add Event Data

The entry point is indicated by a single headed arrow against the first entity to be accessed. Associated with the entry point is the event data. These are attributes which are entered into the update process and will normally consist of a key attribute, together with any attributes which carry updating information.

Once the event data is identified, it is necessary to check that every function of which the event is part either generates the input attributes or receives them as input.

Figures 8.36, 8.37, 8.38 and 8.39 illustrate the completed ECDs for 3-Week Non-Collection, Notification of Reservation, Return of Loan and Notification of Acquisition.

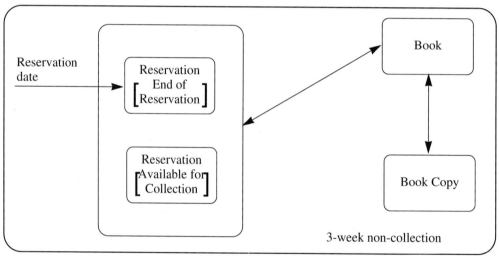

Figure 8.36   3-week non-collection ECD complete

It is usual to position the initial entity to be accessed towards the top of the diagram. Thus some of the entities in the ECDs have been repositioned. In Figure 8.36, the ECD for 3-week non-collection event, the entry point is identified as the 'end of reservation' role of the Reservation entity. It is because the reservation ends through non-collection that the Book Copy can become available for a different reservation.

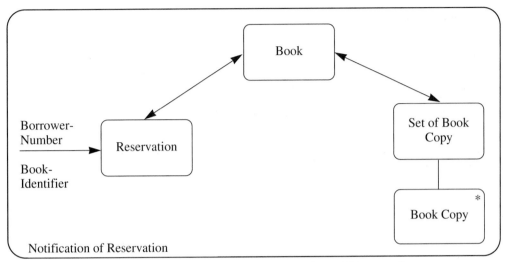

Figure 8.37   Notification of reservation ECD complete

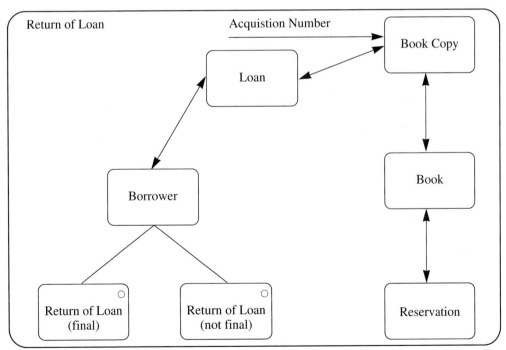

Figure 8.38   Return of loan ECD complete

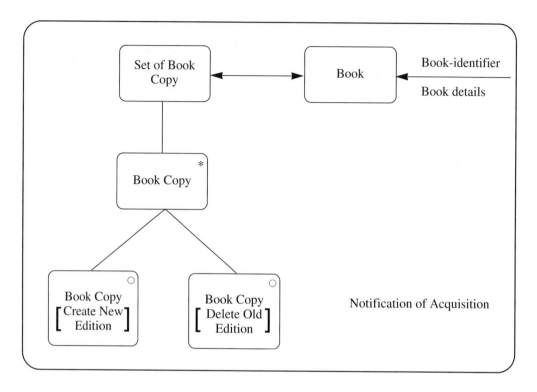

Figure 8.39   Notification of acquisition ECD complete

8.6   DOCUMENTATION

The documentation of the Enquiry Access Paths will include:

- a set of diagrams (see Figure 8.40);
- enquiry trigger attributes (see Figure 8.40);
- enquiry trigger selection criteria documented on the relevant Function Definition Description.

The documentation of the Effect Correspondence Models will include:

- a set of diagrams (see Figure 8.41);
- event data (see Figure 8.41).

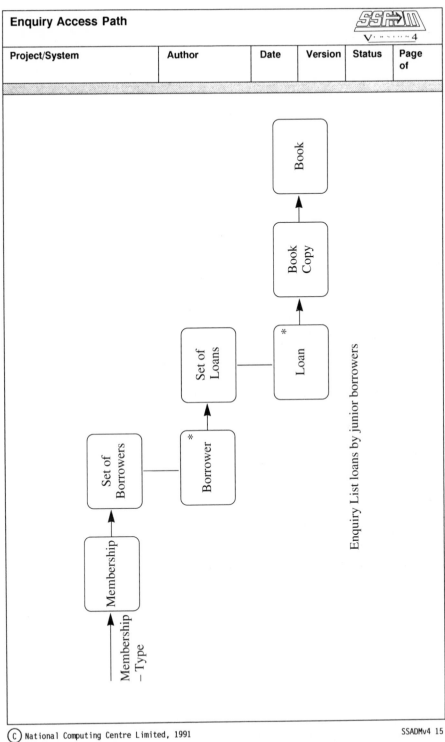

Figure 8.40   Example Enquiry Access Path

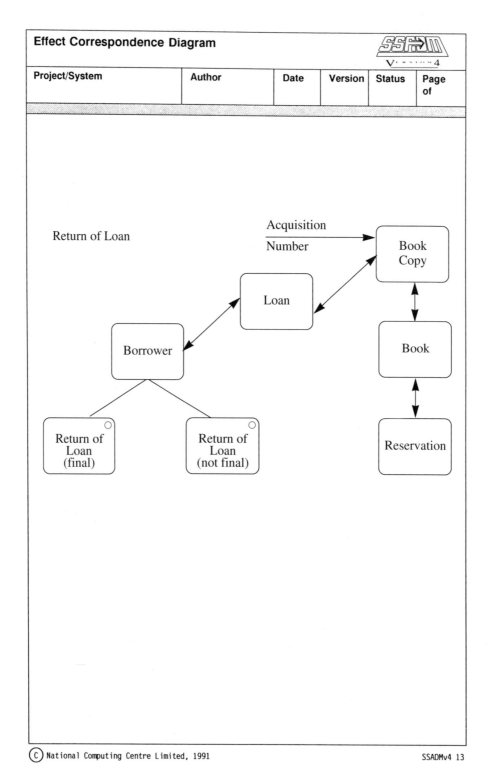

Figure 8.41   Effect Correspondence Diagram

8.7   RELATIONSHIP WITH OTHER MODELS

8.7   Enquiry Access Path

All the components of the LDM are necessary inputs to Enquiry Access Path modelling. The LDS is used as the basis of the models and the correct interpretation of the LDS is checked using the Entity and Relationship Descriptions. The detailed analysis necessary to develop the Enquiry Access Paths enables early validation of the LDS. As has been seen, it may cause changes to the LDM by adding new entities and by changing or adding new relationships.

The I/O Structures provide a definition of the enquiry input and enables identification of the enquiry trigger. The output structure is used to identify the attributes and hence the entities to be accessed to generate the response to the enquiry.

To gain the greatest benefit, in the validation of the Logical Data Model, the Enquiry Access Paths should be completed during the development of the LDM. A copy of the LDS will also be constructed to show all the required entry points.

It is the Enquiry Access Paths which form the basis from which Enquiry Process Models are developed in Logical Database Process Design (see Chapter 9).

8.7.2   Effect Correspondence Diagram

One Effect Correspondence Diagram is drawn for each event identified during ELH analysis. Consequently the ELHs form an important input to the development of ECDs.

The components of the LDM are also an important input to the technique. The LDS is used to ascertain the correspondences between effects and to identify non-update entities which need to be included in the ECDs to enable navigation. The Entity Descriptions are checked to ensure the attributes to be updated have been defined and the event data is validated against the I/O Structure Descriptions.

The ECDs, with the ELHs, are used as input to Logical Database Process Design. Update Process Models are created from the ECDs (see Chapter 10).

8.8   SUMMARY

Enquiry Access Path analysis forms the first steps of Enquiry Process Modelling. It is concerned with:

 – identifying the entities to be accessed to retrieve the enquiry data;

 – identifying the enquiry entry point and any selection criteria;

 – establishing a valid navigation path between the identified entities.

It is important that a clear understanding of the requirements of the enquiry is developed with the users to avoid any misinterpretation of the enquiry output. The analysis of the enquiry access paths also forms a valuable validation technique for the LDM.

The Enquiry Access Path is a new model introduced in SSADM Version 4. It can be regarded as a graphical form of the access detail that was part of the Logical Enquiry Process Outline (LEPO) of SSADM Version 3. However, because Enquiry Access Paths are formally developed at a much earlier stage than LEPOs, they may be used for the essential validation of the Required Logical Data Model as well as providing the basis for the Enquiry Process Models.

Effect Correspondence Diagramming is the second technique of Entity-Event Modelling.

An ECD is concerned with:

- showing which entities are affected by an event:
- establishing how the different effects on the entities are interrelated;
- identifying the update entry point and event data.

It is a useful technique for validating the ELHs and, although completed during Entity-Event Modelling, it constitutes the first steps towards Update Process Modelling.

The Effect Correspondence Diagram is also a new modelling technique introduced in SSADM Version 4. It can be considered as a graphical form of the access detail that was part of the Logical Update Process Outlines (LUPO) of Version 3.

9 Logical Database Process Design

9.1 INTRODUCTION

The techniques described in this chapter aim to translate the models produced during Requirements Specification into a detailed logical specification. This logical specification may in turn be translated into a physical design for the system on a particular software and hardware platform. Logical design permits consideration of design issues uncluttered by implementation aspects, producing a design which optimally satisfies the user requirements. The translation into physical design can thus be treated more objectively and is likely to produce a more efficient, relevant and reliable implementation.

The logical specification also makes the maintenance process easier. It is possible to examine the consequences of proposed enhancements in terms of the logical definition, ensuring that the logical integrity of the system is maintained and avoiding ad hoc and poorly structured changes. This should make it simpler to plan and perform maintenance. However, this tidy delineation between logical and physical design should be viewed pragmatically. If implementation software has been selected early in the development cycle, it may be appropriate to take this into consideration. For instance, the use of a 4GL for implementation may suggest that non-procedural specifications are more relevant.

Logical Process Design comprises both Dialogue Design and Logical Database Process Design. In Logical Database Process Design models are produced to describe both update and enquiry processes. Enquiry Process Models (EPM) are developed from Enquiry Access Paths (EAP) and I/O Structures (produced during logical data modelling). Update Process Models (UPM) are derived from Effect Correspondence Diagrams (ECD) and Entity Life Histories (ELH) (produced during entity-event modelling).

Typically, one model is produced for each function definition although sub-models for interface processes or for processes suitable for re-use may be developed. The processes are defined based upon the logical database that is specified in the Logical Data Model. Ideally, they should be independent of a particular implementation environment.

If the process specifications are to be implemented without using a database management system, then a Process Data Interface may be defined to present the data from the file handling system to the programs in the structure defined by the LDM. Alternatively for performance reasons, it may be necessary to integrate the data access routines and process routines, thus no longer maintaining the direct correlation with the logical specification.

9.2   MODELLING NOTATION

Both the Enquiry Process Model and the Update Process Model utilise the Jackson-like notation of the structure diagram described in Chapter 2. This is appropriate for two reasons. First the concepts of sequence, iteration and selection are fundamental to the description of processes. Secondly, the process model is derived from models already described using this notation. The nodes on the Process Model diagram represent process components or process structures. Process structures show the relationship between the process components.

Update and enquiry processes are not always independent and may be related in two ways:

- an enquiry process may occur as a precursor to an update process;
- an update process which requires an output report may have to be extended using the enquiry process modelling technique.

9.2.1 Enquiry Process Models

When preparing an Enquiry Process Model (EPM) the following procedure should be followed:

- Develop an input data structure from the Enquiry Access Path (EAP). The EAP shows how the enquiry navigates across the database and thus reflects the required input data structure.
- Prepare an output data structure from the I/O Structure.
- Match the corresponding input and output structures to form a process structure.
- Derive appropriate operations from the ELHs and allocate to the process structure diagram.
- Identify conditions and allocate to the process structure diagram.

9.2.2 Update Process Models

An Update Process Model (UPM) is prepared by applying the following procedures:

- Incorporate into the Effect Correspondence Diagram (ECD) for the event entities needed to fulfil any required enquiry output.
- Derive a Jackson-like structure diagram directly from the extended ECD.
- Allocate to the process structure diagram appropriate operations derived from the ELHs.
- Identify and allocate conditions to the process structure diagram.

9.2.1   Matching Data Structures

The central technique in the preparation of an EPM is that of matching the input and output data structures. The technique is derived from a similar procedure used in Jackson Structured Programming. Essentially, the required process structure is one which embodies both input and output data structures, and the process structure is, in fact, determined by the data structures. A correspondence between the input and output structures is shown by double-headed arrows drawn between them.

The correspondences between the structures can be found by applying the following rules (from Burgess 1987):

- The input structure should be considered by starting at the root node and working to the bottom, and working from left to right across each subtree in the structure. This provides a consistent approach for the comparison of the data structures ensuring that all the nodes are considered. In Figure 9.1 the nodes should be considered in the order 1,2,3,4,5 and 6.

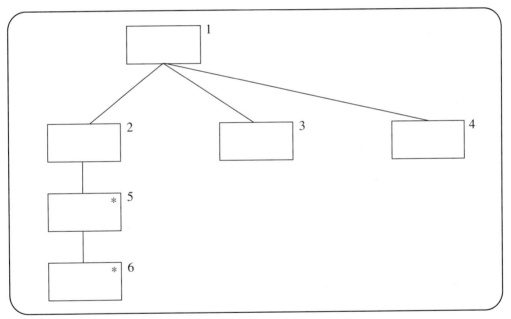

Figure 9.1 Order for considering correspondence

- A component in one data structure may be matched with at most one component in the other data structure.

- Corresponding data components should have the same number of instances in each of their data structures and must occur in the same order. In Figure 9.2 the node 'Order' occurs for every order placed by a customer. There is no node in the output data structure (Figure 9.3) which has the same number of instances and hence 'Order' may not correspond with an output node. If nodes with a different number of instances were allowed to correspond, then the resulting process element would have a different number of inputs and outputs. When the nature of the process suggests that nodes with different numbers of instances correspond and the grouping of the input instances has no direct correlation with the grouping of the output instances, a *boundary clash* is said to occur.

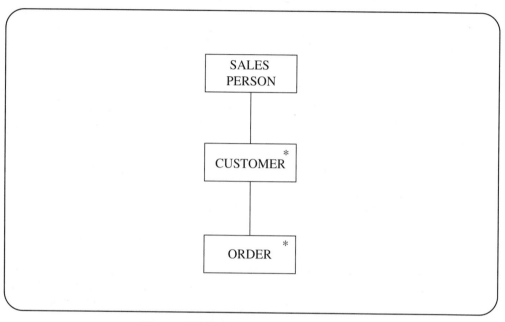

Figure 9.2 Input data structure.

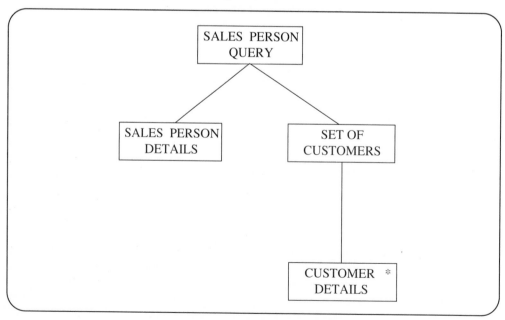

Figure 9.3 Output data structure

– The relative positions of the matched nodes have to be maintained as the structures are considered. Figure 9.4 shows an invalid matching of nodes where their relative positions in the structures have not been maintained.

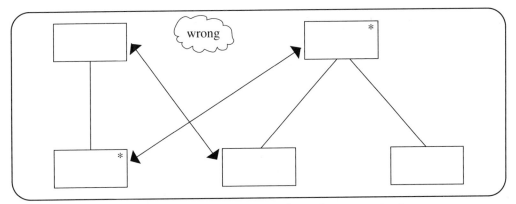

Figure 9.4   Invalid correspondence between stuctures

– When elements of sequences are matched the correspondence must not conflict with the sequence of the individual elements. Figure 9.5 shows an invalid example where the matching conflicts with the sequences in the input and output data structures. When the input and output data structures clash in this way, the required order for the output data does not match the given input order. This is known as an *order clash*.

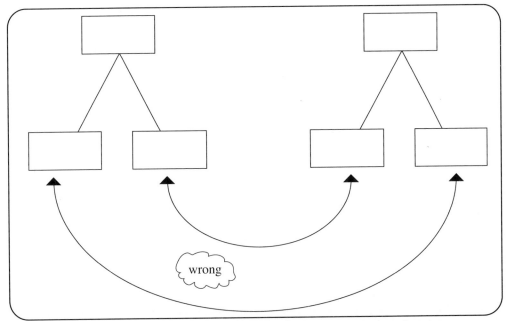

Figure 9.5   Invalid correspondence between sequence structures

– When elements of a selection are matched, the order of the correspondence does not matter (Figure 9.6a) as only one of the selected elements occurs for a particular instance of the data structure. However, it is usual to redraw one of the data structures so that both selections have the nodes in the same order (Figure 9.6b).

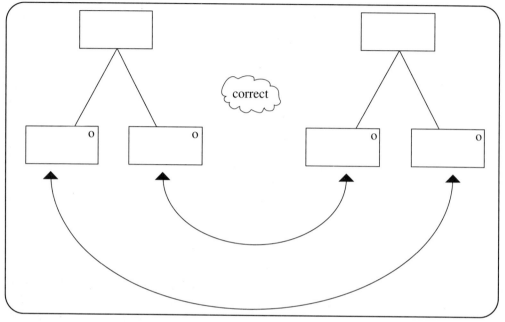

Figure 9.6a   Correspondences for selections

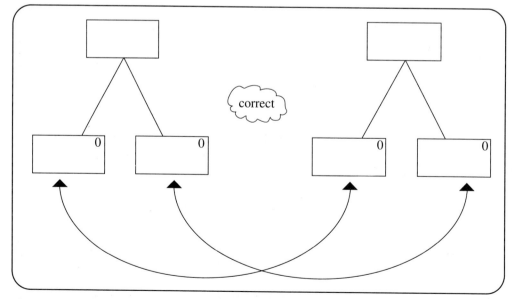

Figure 9.6b   Correspondence for selections – redrawn

Consider the Logical Data Structure shown in Figure 9.7. A report giving the total value of the orders placed by each customer for each salesperson is required. The input and output data structures for the enquiry are given in Figures 9.2 and 9.3 respectively. The required output must detail the total sales for each salesperson.

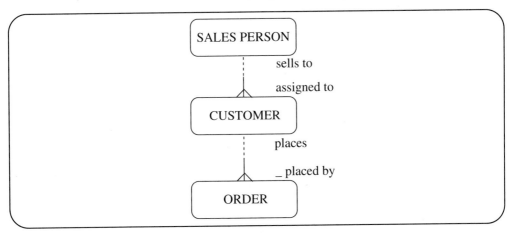

Figure 9.7 Fragment of LDM

In this example the correspondence between the nodes 'Salesperson' and 'Salesperson Query' is straightforward. However, the node 'Order' in the input data structure and the nodes 'Salesperson Details' and 'Set of Customers' in the output data structure do not have corresponding nodes. In Figure 9.8 the correspondences are indicated by double-headed arrows.

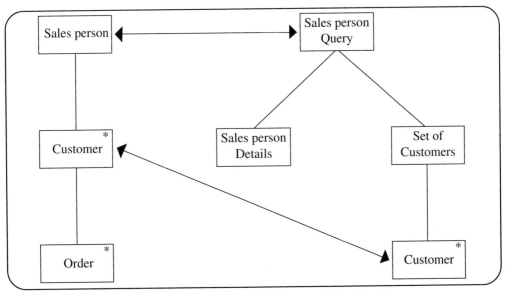

Figure 9.8 Correspondences between data structures

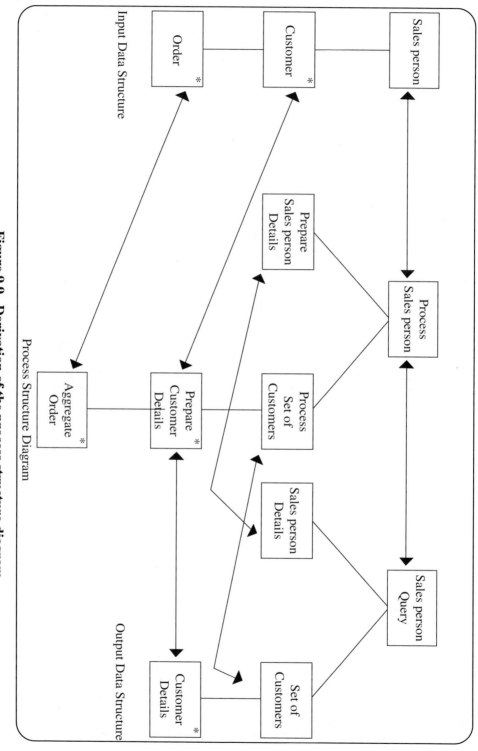

Figure 9.9   Derivation of the process structure diagram

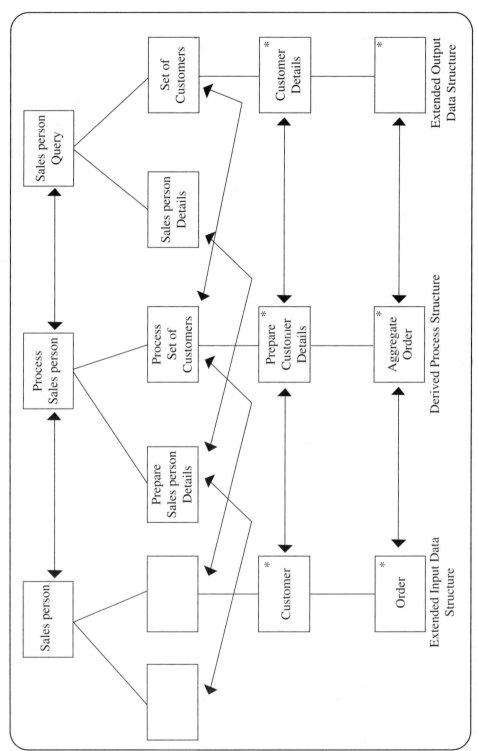

Figure 9.10   Extending the input and output data structures to produce the derived process structure

In order to produce a process structure which will match both input and output data structures, a new structure diagram is produced where each of its nodes matches with one node from the input data structure or one node from the output data structure. Figure 9.9 shows the way in which the input data structure is mapped through this hybrid process structure to the output data structure. The derivation of this hybrid structure can be more easily seen by first extending both input and output data structures with dummy nodes, so that every node in each structure has a corresponding node. This produces two structure diagrams with complementary dummy nodes. The process structure diagram is then produced by superimposing these diagrams upon each other. It is not necessary to draw the extended data structures as shown in Figure 9.10 when preparing a process structure. They are reproduced here merely to illustrate the procedure. However, this diagram illustrates how the derived data structure is designed to encompass both input and output data structures.

9.3   STRUCTURE CLASHES

When the input and output data structures are matched structure clashes may occur. These may be of three types:

- ordering clash;
- boundary clash;
- interleaving clash.

9.3.1   Ordering Clash

This occurs when the sequence of data components is different in the two structures. Consider the enquiry 'List all the orders, in order-number sequence, for a particular salesperson.' The Input Data Structure is shown in Figure 9.2 and the Output Data Structure is given in Figure 9.11. The input structure presents the 'Order' occurrences in order-number within customer-number sequence. The output structure requires the data in strict order-number sequence – an ordering clash. An ordering clash may be resolved by:

- modifying the LDM so that the input data structure may be changed;
- specifying an additional process (a sort) to resequence either the input or the output data.

The first option could be achieved by adding a relationship from 'Salesperson' to 'Order'. However, changing the LDM may have an undesirable effect upon other processes which it must support and great care should be taken. The separation of logical and physical design issues is best maintained by leaving the optimisation of the processes to physical design and only ensuring at this stage that they are logically supported by the LDM. Hence, the second option may be more appropriate. An intermediate output data structure could be defined so that it is compatible with the input data structure derived from the LDM. An additional interface process is then defined which reorders the data to satisfy the output requirement. This also ensures that the process models are a succinct description of the logical requirements with the necessary implementation detail only described in the interface processes.

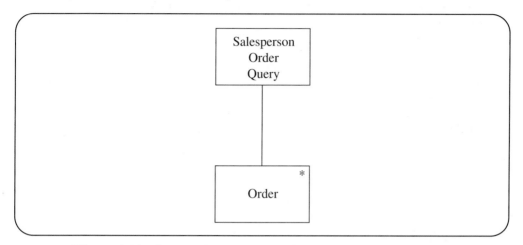

Figure 9.11   Output data structure for enquiry 'List all orders'

9.3.2   Boundary Clash

This typically occurs when the grouping of data elements in the LDM as reflected in the input data structure is different from that required by the output mechanism. For instance, a process is required to display the orders placed by a customer. One or more items may be ordered on each order but the VDU can only display a maximum of 24 lines of text at any one time. The first order may require 10 display lines and the second 18 display lines. Thus the end of the second order does not coincide with the boundary dictated by the screen. This incompatibility of the input grouping of order items (dictated by the master entity Order) and the required output grouping (dictated by the VDU) is a boundary clash. This problem may be eliminated if a report or application generator is used which automatically handles the grouping of the output data. Alternatively, a separate process may be defined to handle the formatting of the output.

Again, as with the Ordering clash, it is simplest to define an intermediate output data structure, so that it is compatible with the input data structure, and then define an additional output process to perform the necessary formatting. In other circumstances, an additional input process may have to be defined to overcome the boundary clash when the grouping of the input data is not the same as the logical grouping required by the logical process.

9.3.3   Interleaving Clash

An interleaving clash occurs when the events for one entity are interleaved with the events for another entity. This results in a data structure whose elements alternately affect different entities. Thus, after a database effect has been applied to one entity occurrence, its state must be stored whilst the next element in the input data structure is applied to another entity occurrence. A common example of this occurs when a database process has to navigate along a recursive relationship such as a parts explosion where the data for several different entities has to be available to the process. An interleaving clash can be resolved by defining the process so that it temporarily stores entity occurrences whilst the other entity occurrence is processed.

9.4   ALLOCATION OF OPERATIONS

The specific types of operation that may be allocated during update and enquiry process modelling are as follows:

- read <entity type> by key
- define set of <entity type> matching input data
- read next <entity type> in set
- read next <detail> of <master> [via <relationship>]
- read <master> of <detail> [via <relationship>]
- invoke <common process>
- fail if state indicator of <entity> outside <value range>

Other operations which have been specified during entity life history analysis should also be allocated to the structure diagram. These include:

- Store <attribute>
- Store keys
- Store remaining attributes
- Store <attribute> using <expression>
- Replace <attribute>
- Replace <attribute> using <expression>
- Tie to <entity>
- Cut from <entity>
- Gain <entity>
- Lose <entity>

The last four operations in the list may not be relevant if a Relational Database Management System is used for implementation because relationships are implemented as foreign keys.

9.5   LOGICAL DESIGN ISSUES

There are several issues which need to be addressed during Logical Design:

- specification of Logical Success Units;

- identification of Common Processing;

- resolution of Recognition Problems;

- database integrity.

9.5.1 Logical Success Units

A Logical Success Unit (LSU) is a set of processing which must succeed or fail as a whole. In database terms, it is a transaction which must be applied successfully to the database and committed or must be rolled back. Whilst an LSU is being executed the data which it accesses is locked so that no other process can modify it. It is only within the scope of a success unit that the consistency of the data can be guaranteed.

In Logical Database Process Design (LDPD) each update process is regarded as an LSU as it is triggered by an event which itself must succeed or fail as a whole. The status of enquiry processes is less well defined. The simplest solution is to also view each enquiry process as an LSU. However, this is not always practical as it would result in the database being locked during each enquiry, so preventing any concurrent update processes. In general, it is preferable to leave detailed consideration of such operational issues for Physical Process Specification when Physical Success Units are defined. However, there are occasions when it is essential that the data retrieved is wholly consistent with the database and this justifies identifying the enquiry process as an LSU. For example, an on-line order enquiry which displays details of the order lines would sensibly be designated as an LSU preventing modification of any of the data associated with the order. Equally a payroll report giving the pay details of employees might be designated an LSU. However, time-consuming enquiries like this are likely to be processed off-line when there is no update contention.

If large enquiries are to be handled on-line, they can be split into smaller sub-enquiries each of which is treated as an LSU. For example, if all the patients assigned to a particular consultant are being retrieved, they can be presented in groups of ten. The only sound justification for sub-dividing an enquiry in this way during logical design is to give the user the opportunity to abandon the query before all the patient records have been read. The sub-divided enquiry only locks ten records at a time. Consequently patient details which have been displayed and are subsequently no longer locked, may be modified, reducing the validity of the enquiry as a whole.

9.5.2 Common Processing

An important objective of logical design is to minimise duplicated effort during physical design. This requires the identification of elements of processing which are common to more than one logical database process. Ideally these elements of common processing are specified once and then constructed once. The advantages are obvious:

- reduction in effort in construction;

- fewer opportunities for errors;

- easier maintenance as there is only one process to modify.

It is difficult to prescribe a procedure which will unerringly pinpoint elements of common processing – the experience of the developer is still the major factor. However, when two or more logical database processes share common processing, a separate process model may be defined for this common functionality. Ultimately, this becomes a common module called by each process which requires this processing. Some elements of common processing are more easily identified during physical design and this is considered further in Chapter 11 which deals with Physical Process Design.

9.5.3   Recognition Problems

On occasions it is not possible to test a condition at the most appropriate point during a process because the necessary data is not available. This problem may be overcome by using one or more of the following techniques:

- Store an additional (derived) data item in anticipation of the test. For instance, sales orders may only be placed by a customer if the value of outstanding orders is less than a preset amount. The update process which inserts a new order must either read all the outstanding orders to make this test or access an additional derived data item 'outstanding order value'. The latter is likely to be the most efficient solution.

- Introduce a 'multiple read ahead' so that the condition can be checked. For example, an update program reads a transaction file containing records batched in groups of five. Each batch is preceded by a batch header record which contains batch totals for the numeric fields. If there is a discrepancy between the batch total and the actual batch, an error should be flagged and no update should occur. In this circumstance a 'multiple read ahead' strategy reads the batch header record and all five transaction records before the batch is processed.

- Design two processes and so pass through the data twice. The first process is designed to identify and mark the necessary check points for the second process.

- Apply a backtracking technique. This involves assuming that a particular condition is true, and continuing the processing based upon this assumption. If the assumption is proved false the process backtracks, undoing any harmful effects of the processing. An alternative assumption is then made and processing continued.

9.5.4   Database Integrity

It is important that the integrity of the database structure is maintained by the processes that operate upon it. Ensuring consistency between master and detail records is one form of integrity that is common to all applications (also known as referential integrity, for further reading see Date 1991). For instance, a master entity should not be deleted whilst there are still detail entities linked to it in the database. Other integrity rules are application specific and relate to the business rules for the enterprise. For instance, a borrower may not be allowed to borrow books whilst the borrower has unpaid fines. Such integrity constraints will have been identified during entity-event analysis and reflected in the state indicators in the Entity Life Histories.

Appropriate statements must be included in the logical specification of each process which will enforce the integrity constraints and which will cause the process to fail if the database integrity is not maintained. For each integrity error suitable error messages should be defined. As these error conditions are a consequence of the logical design, it is important that they are addressed at this stage. Equally, the requirement of usability necessitates that the user is directly involved in the definition of the error handling and error handling messages. It is a common misconception of system developers that error handling is less important because error detection is a small part of system activity. The user's inability to respond correctly when errors have been detected results in perpetuating misuse of the system and generating user dissatisfaction.

9.6　CONSTRUCTION OF PATIENT EXAMPLE

A patient database in a general hospital will be considered in order to describe the steps involved in preparing an EPM. The relevant fragment of the Logical Data Model is shown in Figure 9.12. The proposed system will hold information concerning the patients who have been admitted to hospital and record whether they have been admitted to a general ward or a maternity ward. The system will also record the treatments administered to the patients during each admission.

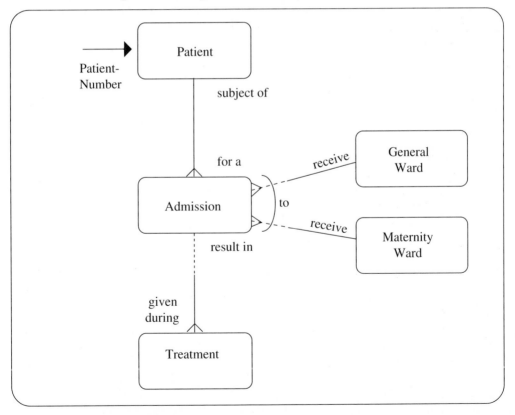

Figure 9.12　Required view of logical data model: patient database

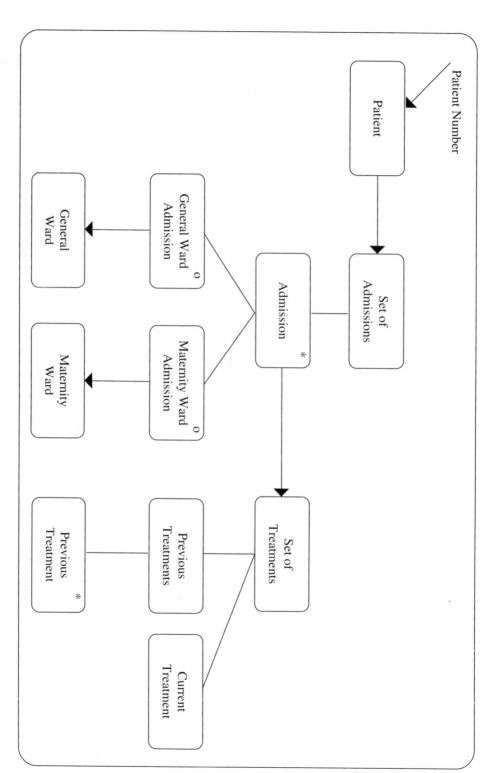

Figure 9.13　Enquiry access path: patient query

9.6.1   Preparing an Enquiry Process Model

During the analysis phase a Function Definition for the following enquiry was specified:

> 'For a given patient, list all admissions and for each admission specify whether it was general medical or maternity and the last treatment recorded.'

The basic steps for constructing an Enquiry Process Model are described below:

A: *Specify the Enquiry Name*

The unique enquiry name is specified in the Function Definition and the Physical Process Specification. For instance, the enquiry in the example is named 'Patient Query'.

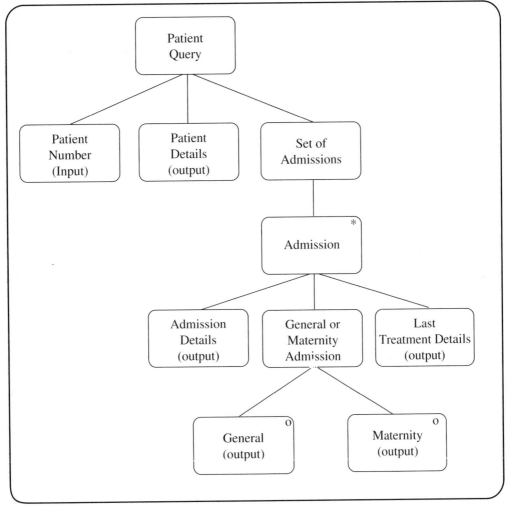

Figure 9.14   I/O structure: patient query

B: Specify the Enquiry Trigger

The enquiry trigger is normally defined on the Enquiry Access Path and comprises the data items input to enable the enquiry process to be performed. Typically, this is the key of the entity used as the entry point of the enquiry and possibly some further selection criteria. In Figure 9.12 the enquiry trigger is shown on the LDM as patient-number.

C: Specify the Enquiry Access Path

This activity has been described earlier in Chapter 8. The EAP is shown in Figure 9.13.

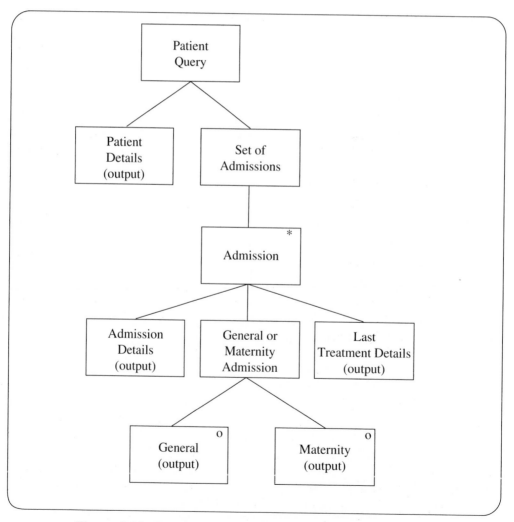

Figure 9.15 Enquiry output data structure: patient query

D: Specify Enquiry Output

The enquiry output comprises the data items output by the enquiry excluding error conditions or messages. A Jackson-like structure is normally used to describe the enquiry output, although if the structure contains no repeating groups a list of data items may be sufficient.

The enquiry output structure is developed from the I/O Structure by simply omitting those parts of the structure which describe input from the user. Figure 9.14 shows the I/O Structure and Figure 9.15 gives the resulting enquiry output structure. Note that 'hard boxes' are used for this diagram to help distinguish it from the I/O Structure.

E: Group accesses on the Enquiry Access Path

Two steps are involved in preparing the input data structure. The first requires grouping accesses that are in one-to-one correspondence together. Ultimately these will be handled by the same processing element (and possibly its child nodes) in the EPM. This produces the diagram shown in Figure 9.16.

F: Convert to Jackson-like Structure

The Enquiry Access Path diagram with grouped accesses is converted into a Jackson-like structure. Each grouped access becomes a node in the new diagram for which additional nodes may have to be created so that the diagram is syntactically correct.

The diagram in Figure 9.17 represents the input data structure for the database process and reflects the structure of the LDM. It is important to appreciate that each node, not just elementary leaf nodes, on the diagram may represent an elementary or grouped access to the logical database. In other uses of the structure diagram, the elementary leaf nodes represent the actual components of the structure whilst the intermediate nodes reflect the structure. For instance, in Figure 9.17 the node 'Patient' represents the access to the patient entity and the set of accesses to the 'Admission' entity. These latter accesses are individually represented by the 'Admission' node. However, as 'Admission' in the original EAP is a selection and 'Set of Treatments' is an iteration, they cannot be combined directly. A new child node 'Admission Type' is introduced for the selection.

G: Identify Correspondences Between Input and Output Data Structures

The correspondences between the structures are identified by linking nodes on the corresponding diagrams which relate to the same group of data items. These corresponding nodes are linked by double-headed arrows as shown in Figure 9.18. The highest level nodes 'Patient' and 'Patient Query' are in correspondence as they both refer to logically the same group of data items. Similarly, the nodes 'Admission' in both structures also correspond (the names of the nodes can be a useful guide). However, there is no corresponding output node for the input node 'Previous Treatment' and no input node for the output node 'Admission Details'. As described earlier, structure clashes become evident at this stage. A requirement to page the output would result in boundary clash as the

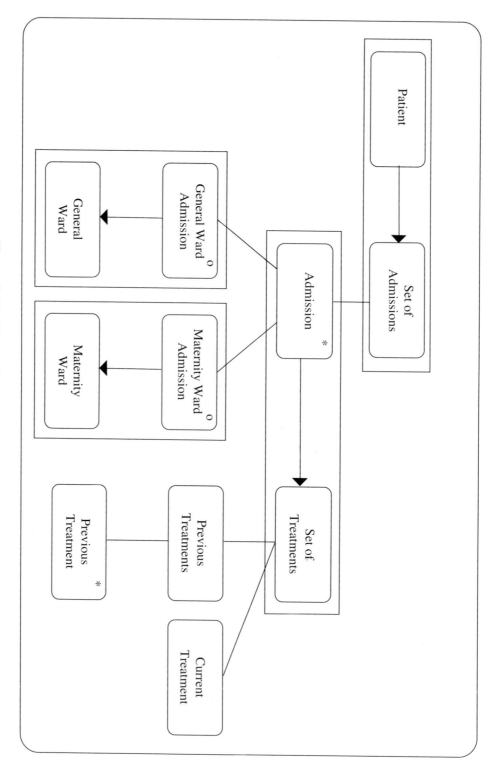

Figure 9.16   Grouped accesses: patient query

number of admissions per patient would not necessarily exactly fill a page. This structure clash could be resolved during Physical Process Specification by specifying an additional output process and need not be considered at this stage.

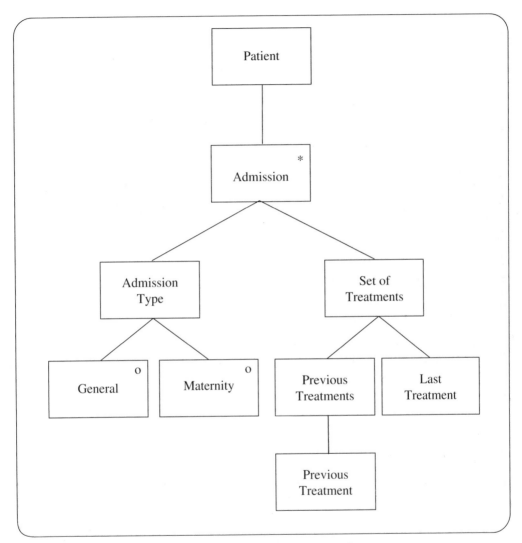

Figure 9.17 Input data structure: patient query

H: Match the Input and Output Data Structures

The Process Structure Diagram is derived by this procedure and is shown in Figure 9.19. There are nodes for the matched nodes in the input and output data structures and also for those which could not be matched such as 'Previous Treatment'.

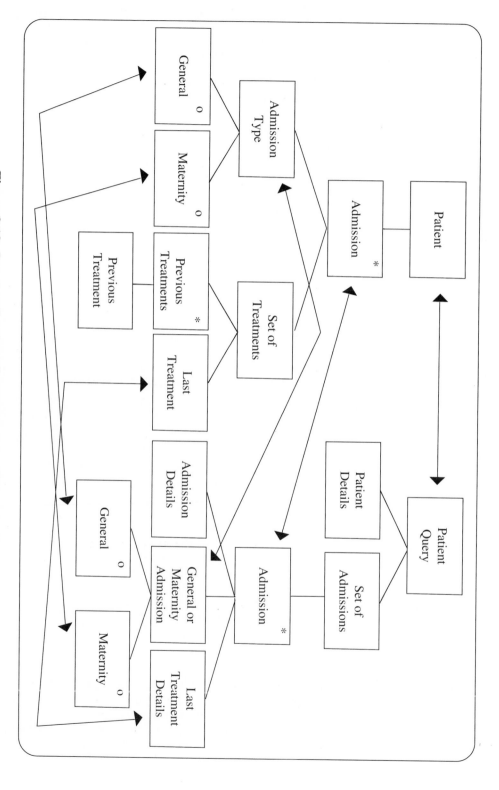

Figure 9.18   Input and output data structures with correspondences

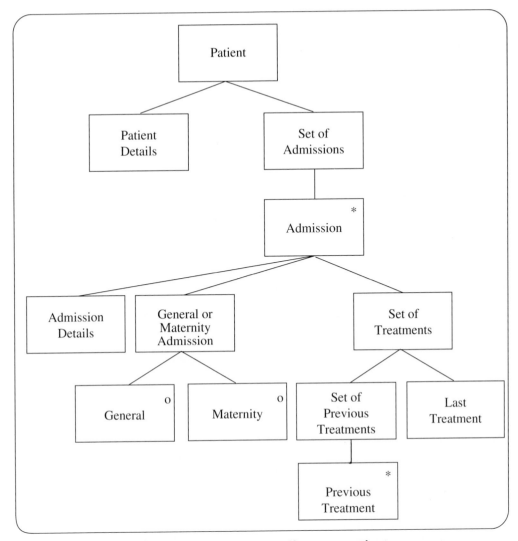

Figure 9.19 Process structure diagram: patient query

I: List the Operations and Allocate them to the Structure

At this stage only read operations are allocated. Other operations are allocated during Physical Design but may be specified non-procedurally at this point. Non-procedural specification is preferred to avoid biasing or prejudicing the implementation. The operations are derived from the relevant nodes on the ELHs. Figure 9.20 shows the Process Structure diagram with operations added.

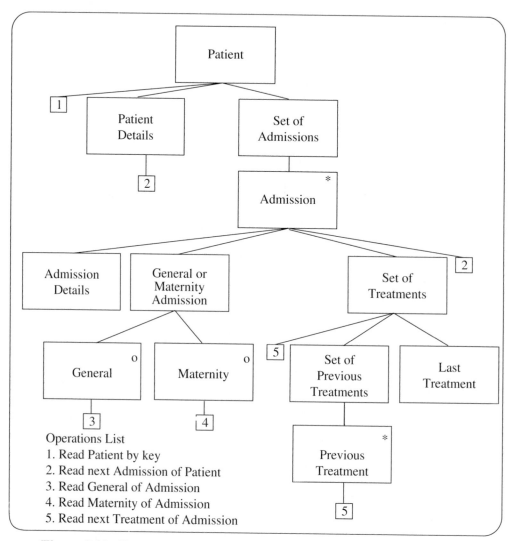

Figure 9.20   Process structure diagram with operations: patient query

J: Allocate Conditions to the Structure

The selection and iteration constructs of the structure diagram represent optional processing components. Conditions should be specified to determine when these processing elements are invoked. Typically, the conditions associated with selection constructs test the value of an entity's state indicator whilst those associated with iteration constructs test for 'end-of-data-set'. Figure 9.21 shows the process structure diagram with conditions added. Note that the first 'Admission' occurrence is read before entering the iteration and subsequent reads are performed after each iteration.

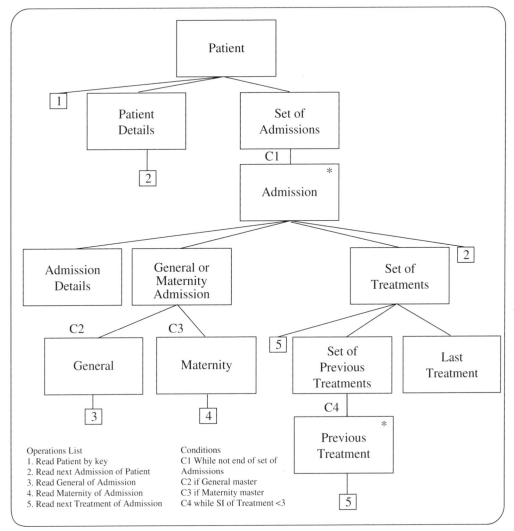

Figure 9.21 Process structure diagram with conditions: patient query

In Figure 9.21 the conditions are coded and listed at the bottom of the diagram. It is acceptable to write the actual condition above the iterated or selected node.

K: Specify Integrity Error Conditions

Ensuring that the integrity of the database is maintained is largely the problem of Update Processes. However, it is important that Enquiry Processes only produce information from a logical database whose integrity has been maintained.

The most obvious source of failure for an enquiry is when the first or subsequent access points to an entity are missing. Other integrity errors may occur when the entity occurrence accessed does not have a valid state indicator value for the enquiry. The valid state indicator values are specified in terms of the valid prior state indicator values assigned during the preparation of the ELHs. Each fail condition must be specified in the operations list and added to the structure diagram as an operation immediately after the relevant read operation. This documents the way the process will handle integrity errors.

Note that any error handling operations that may have been identified during dialogue design should be transferred to the process diagram.

L: Specify Error Outputs (logical and physical)

Whenever a process encounters an integrity error, a suitable error message must be produced. This may be incorporated with the valid output or produced on a separate report. In either case the message must enable the user to understand why the process has failed and provide sufficient information so that the user can initiate action to remedy the problem.

M: Walkthrough the Structure

It is important that the Process Model is validated via some quality procedure such as a structured walkthrough. This will ensure that:

- the syntax is correct;

- the process model satisfies the functional requirement;

- the error handling procedures are appropriate.

9.6.2   Preparing an Update Process Model

An Update Process Model (UPM) is prepared for each event that was identified during entity life history analysis. The Patient Database again provides the example. The preparation of an UPM is similar to that of an EPM. Only those steps in the procedure which are different will be described in detail. The event considered here is:

'Record the details of a new treatment for an existing patient.'

This Update Process operates upon the Logical Data Model shown in Figure 9.22.

An Update Process Model is prepared by following steps A to M. The first three steps are performed during entity-event modelling and ELH analysis.

A: Specify Event Name

B: Specify Event Data

C: Specify Effect Correspondence Diagram

The event name uniquely identifies the event in other models and documents (ELHs, ECDs, Function Definitions and Physical Process Specification). The event data (ie event trigger) is usually documented in the Function Definition. In this example this includes the Admission-Number, the Patient Name to verify that the correct patient is being treated and details of the treatment prescribed. The ECD is used to validate the set of ELHs and represents the way an event effects the logical database. Figure 9.23 shows the ECD for the event 'Patient Treatment'.

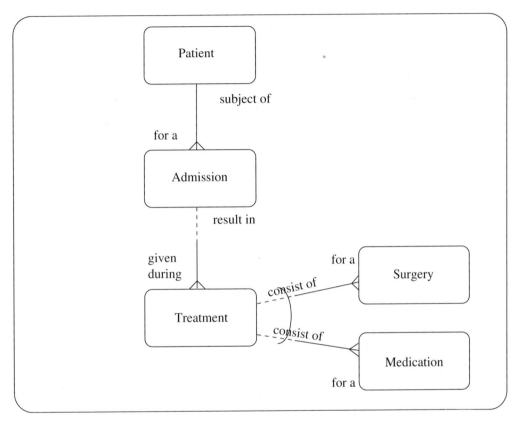

Figure 9.22 Logical data model: patient treatment

D: Specify Event Output

If the event output data is not merely 'Event completed OK' then it should be described in the same way as 'enquiry output' for an enquiry process.

E: Extend the ECD with Enquiry-only Entities

If the event is coupled with an enquiry, it may be necessary to extend the ECD so that it shows the accesses to the enquiry-only entities.

F: Group the Effects in one-to-one Correspondence

The effects which are in one-to-one correspondence are grouped together by drawing a box around them. Each of these groups and each of the ungrouped elementary processes are then named to reflect the processing being performed. In Figure 9.24 the grouped effects for Patient, Admission and Treatment are named Patient Treatment.

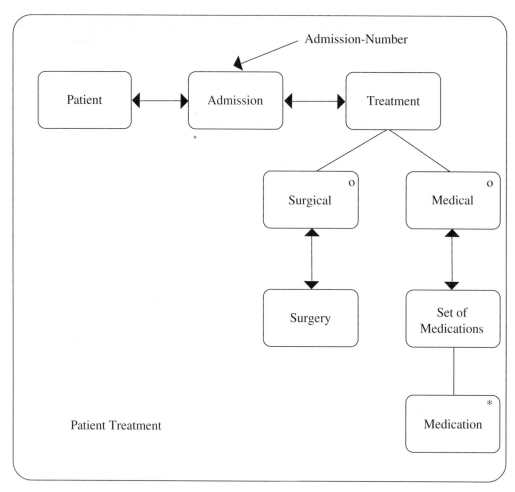

Figure 9.23   Effect correspondence diagram: patient treatment

The next three steps are largely the same as their counterparts in Enquiry Process Modelling.

G: List Operations

When preparing the list of operations it is convenient to group by entity in the following sequence:

1 –   an operation to read the entity;

2 –   operation(s) to raise error(s);

3 –   an operation to create the entity;

4 –   operations from the relevant effect on the ELH's ignoring 'Gain' and 'Lose' operations if they have been used as they are not appropriate for current DBMS products;

5 – an operation to set the entity's State Indicator;

6 – an operation to write the entity.

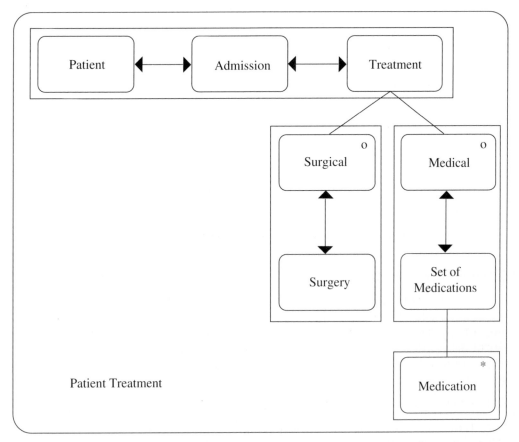

Figure 9.24   ECD with grouped effects and named processes: patient treatment

H: Convert to a Jackson-like Structure

The process structure diagram is shown in Figure 9.25. Note that each of the named groups and each of the named ungrouped elements from the ECD in Figure 9.24 has been translated into a node on the process structure. The grouped nodes such as 'Patient Treatment' handle the accesses or effects for all the entities in the group.

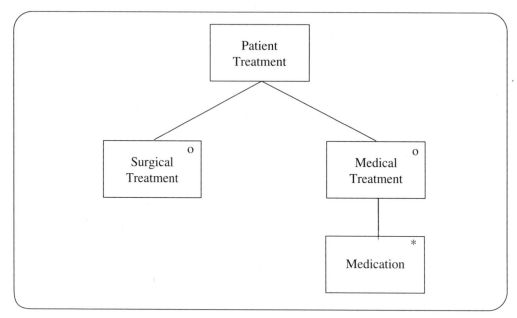

Figure 9.25   Process structure: patient treatment

I: Allocate Operations to Structure

If the operations have already been listed by entity this step is straightforward. The resulting diagram (Figure 9.26(a)) can be rather daunting, though it should be noted that this example affects several database entities and is quite complex. To simplify the document the operations may be listed on a separate sheet (Figure 9.26(b)). As with the preparation of UPMs, the resolution of recognition problems may be specified at this stage.

J: Allocate Conditions to Structure

Conditions are allocated for each selection and iteration in the structure. Figure 9.27 shows the process structure with the conditions added.

K: Specify Integrity Error Conditions

Most of the integrity errors can be derived from the entity life history analysis. However, some integrity constraints may only emerge during logical database process design. Frequently these may be of the form of checks between data items and between entities. For instance, if the number of books borrowed plus the number of books about to be borrowed exceeds the borrower limit, then the update process should fail. Integrity constraints such as these may have been documented in the Required System LDM as domain rules.

The last two steps are the same as in the preparation of Enquiry Process Models.

L: Specify Error Outputs (logical and physical)

M: Walkthrough the Structure

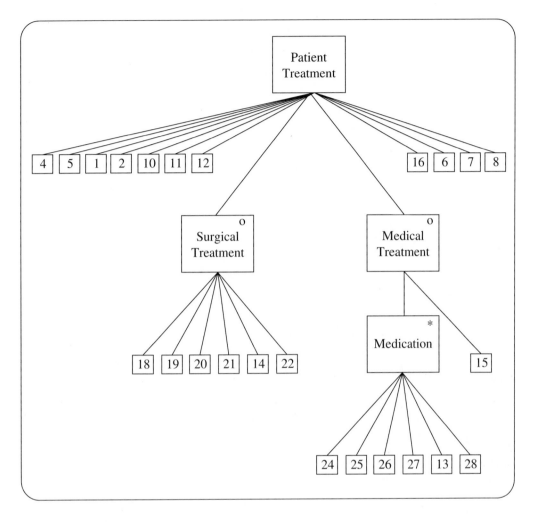

Figure 9.26 (a)   Process structure with operations allocated: patient treatment

9.7   TEXT SPECIFICATION

The specification of the Update Process expressed in the process structure (annotated with operations and conditions) may be described textually. Figure 9.28 shows an equivalent specification prepared in the format of an action diagram. Textual specifications may be useful when verifying the logical processes with users. Equally, when processes are to be implemented using a 4GL they may be a more appropriate method of specification.

Operations List for Patient Treatment
1 Read Patient by key
2 Fail if SI <> 2 or 3
3
4 Read Admission by key
5 Fail if SI <> 2 or 3
6 Replace No-of-Treatments of Admission
   using (No-of-Treatments of Admission + 1)
7 Set SI of Admission to 3
8 Write Admission
9
10 Create Treatment
11 Store keys of Treatment
12 Store remaining attributes of Treatment
13 Replace No-of-Medications of Treatment
   using (No-of-Medications of Treatment + 1)
14 Set SI of Treatment to 1
15 Set SI of Treatment to 2
16 Write Treatment
17
18 Create Surgical-Treatment
19 Store keys of Surgical-Treatment
20 Store remaining attributes of Surgical-Treatment
21 Set SI of Surgical-Treatment to 1
22 Write Surgical-Treatment
23
24 Create Medication
25 Store keys of Medication
26 Store remaining attributes of Medication
27 Set SI of Medication to 1
28 Write Medication

Figure 9.26(b)   Operations list: patient treatment

9.8   SUMMARY

LDPD draws upon many of the earlier products of analysis and design. Enquiry Process Models are prepared from the Enquiry Access Paths and the I/O Structures, whilst Update Process Models are developed from Effect Correspondence Diagrams. This is illustrated in Figure 9.29. The logical processes can be described using structure diagrams or some form of textual specification. This use of the structure diagram is a significant departure from SSADM Version 3 which based its process specification on the Logical Update Process Outline or the Logical Enquiry Process Outline. These were form-based techniques which had no graphical equivalent (though a Logical Access Map could be used in their preparation) and did not support (nor specifically hinder) automatic code generation.

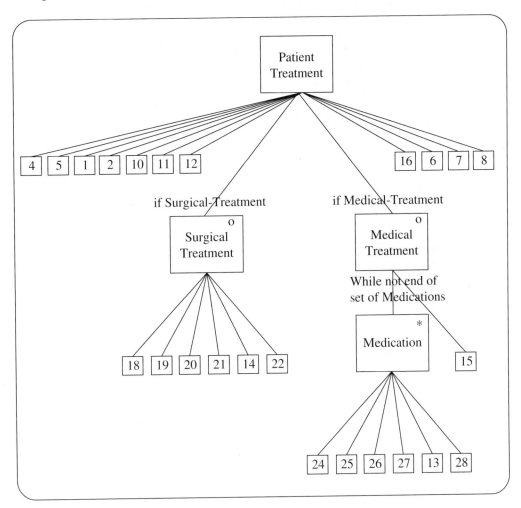

Figure 9.27   Process structure with conditions allocated: patient treatment

The development of process models involves several important issues:

- the identification and resolution of structure clashes;
- the identification of common processing;
- the identification and resolution of recognition problems;
- the specification of integrity error handling.

The consideration of these issues may require the review of previously produced models and, where appropriate, direct user involvement to ensure that the system satisfies the users' requirements.

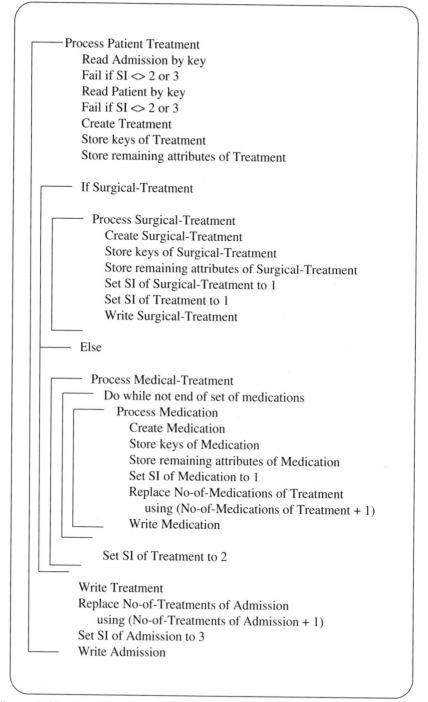

Process Patient Treatment
  Read Admission by key
  Fail if SI <> 2 or 3
  Read Patient by key
  Fail if SI <> 2 or 3
  Create Treatment
  Store keys of Treatment
  Store remaining attributes of Treatment

  If Surgical-Treatment

    Process Surgical-Treatment
      Create Surgical-Treatment
      Store keys of Surgical-Treatment
      Store remaining attributes of Surgical-Treatment
      Set SI of Surgical-Treatment to 1
      Set SI of Treatment to 1
      Write Surgical-Treatment

  Else

    Process Medical-Treatment
      Do while not end of set of medications
        Process Medication
          Create Medication
          Store keys of Medication
          Store remaining attributes of Medication
          Set SI of Medication to 1
          Replace No-of-Medications of Treatment
            using (No-of-Medications of Treatment + 1)
          Write Medication

      Set SI of Treatment to 2

  Write Treatment
  Replace No-of-Treatments of Admission
    using (No-of-Treatments of Admission + 1)
  Set SI of Admission to 3
  Write Admission

Figure 9.28   Text Specification using Action Diagram: patient treatment

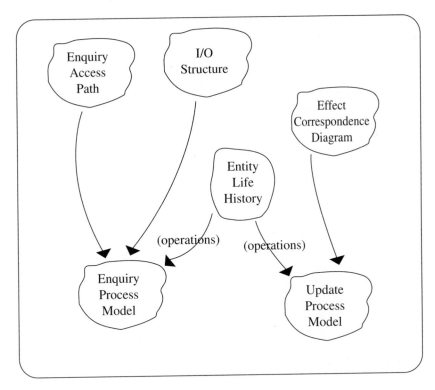

Figure 9.29 Relationship with other models

10   Physical Data Design

10.1   INTRODUCTION

The objective of physical design is to define the physical data structure, the physical processes that operate upon this data structure, and the inputs and outputs of these processes. This specification should utilise the chosen implementation environment efficiently and effectively and adhere to installation standards. The variation in potential implementation environments necessarily means that the approach to Physical Design cannot be tightly prescribed. Different criteria may apply for different styles of implementation (3GL or 4GL for instance) and also for different implementation systems (CODASYL or Relational Database Management Systems). SSADM provides a pragmatic general approach to Physical Design and identifies key principles around which a coherent and specific design approach can be built for particular implementation platforms.

Physical Design maps the Logical Design (the inputs to Physical Design comprise most of the models of Logical Design) onto a particular combination of hardware and software. As such it does not change the Logical Design. However, if during Physical Design the Logical Design is found to be incorrect or incomplete, the Logical Design must be amended. The users should be involved in Physical Design as much as possible and practical.

Physical Design involves the following activities:

- preparation for physical design. This involves the development of the Physical Design Strategy;

- completion of the specification of the functions;

- incremental development of the data and process designs. This demands the iterative performance of the following steps:

    - design;

    - testing against objectives;

    - optimisation;

    - review.

The Physical Design Strategy (PDS) is a component of the Application Development Standards (Figure 10.1) which are derived from the Installation Development Standards and guide the application development process. The development of the Application Development Standards to produce an environment-specific approach is central to the customisation of the general principles, identified in SSADM.

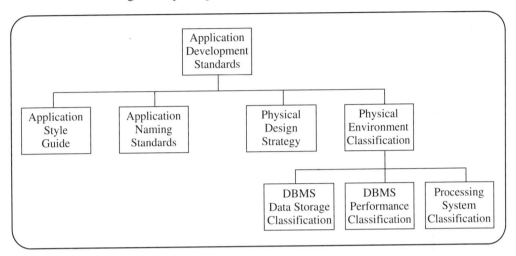

Figure 10.1   Application development standards

Physical Design comprises two inter-related tasks – Physical Data Design and Physical Process Design. Physical Data Design is considered in this chapter and Physical Process Design in Chapter 11. Physical Data Design initially produces a first-cut data design by applying general rules of thumb which are not product specific. This general first-cut data design is then transformed into a product specific design.

10.2   PREPARING FOR PHYSICAL DATA DESIGN

The preparation for Physical Design need only be performed when a new or modified implementation environment is being used for the first time, or perhaps with a new type of application. However, it is useful to review the components of the Application Development Standards relevant to Physical Design at the beginning of the design phase. Preparation for Physical Design involves two activities:

– classifying the DBMS;

– creating the Physical Design Strategy.

10.2.1   Classify the DBMS

Upon completion of the Physical Environment Specification, detailing the implementation environment, the data aspects of the Physical Environment Classification should be completed. Although individual designers may have detailed knowledge of the hardware and software it is useful to document this knowledge, particularly in a large development department where there may be several possible development and operational environments. This helps the less experienced designers in the selection of suitable environments and aids their development and maintenance work. Standard criteria should be used to classify the DBMS. The product vendor may provide the necessary information in suitable format but, if they do not, then the DBMS should be classified using standard forms. Figures 10.2 and 10.3 show typical DBMS. DBMS Data Storage Classification and Performance Classification Forms respectively and comprise the data design elements of the Physical Environment Classification. The forms are illustrated for a typical RDBMS. Once the DBMS has been classified it is possible to prepare space and timing estimation forms. There are no SSADM standards for these forms, although a sample timing estimation document is shown in Figure 10.4. The design of local forms suitable for environment classification also forms part of the definition of Application Development Standards.

DBMS Data Storage Classification				SSADM Version 4		
Project/System	Author	Date	Version	Status	Page of	

DBMS/file handler **RDBMS**

Relationship representation

Table	List	Phantom
Indexes	No	Serial Searches

Amalgamation of entity and relationship data

None	**YES**
Relationship and master	**NO**
Relationship and detail	**NO**
Relationship with master and detail	**NO**
Relationship and relationship	**NO**

Key representation in relationship (logical or physical)

Master to detail/detail to next detail	Physical (indexes)
Detail to master	Logical

Retrieval by logical key

Search	**YES**
Indexing	**YES**
Hashing	**NO**

Implementation of place-near logic

Record clustering

Significant restrictions

SSADMv4 6

Figure 10.2   Data Storage Classification

DBMS Performance Classification					SSADM VERSION 4
Project/System	Author	Date	Version	Status	Page of

DBMS/file handler   **RD BMS**

Transaction logging overhead

Transactions logged for on-line events

Commit/backout overhead

Before and after imaging available

Space management overhead

Compresses null values

Dialogue context save/restore overhead

Standard timing factors

Disc operation:	Time:	Comment:
Read: 30 MS		
Write: 30 MS		
Overflow overhead:		

DBMS operation	DBMS processor time	TP monitor processor time
All reads	4 ms	2 ms ⎫ if disk I/O
All others	4 ms	4 ms ⎭ caused else 0

Performance parameters for available sort packages

Ⓒ National Computing Centre Limited, 1991

SSADMv4 7

Figure 10.3   DBMS Performance Classification

TRANSACTION TYPE:			ON-LINE	BATCH	No OF TRANSACTIONS:			PER:
				DISC ACCESSES		CPU TIME		
RECORD TYPE	NUMBER ACCESSED	ACCESS TYPE	ACCESS PATH	INDEX ACCESS	DATA ACCESSES	DBMS CPU TIME	TP MONITOR CPU TIME	APPLICATION CPU TIME

ACCESS TYPES R = Read S = Store L = Link M = Modify D = Delete U = Unlink

Figure 10.4 Sample timing estimation form

10.2.2   Create the Physical Design Strategy (PDS)

When the design team is familiar with the implementation environment, the design strategy should be specified. This strategy may apply to all similar applications using this environment. The PDS attempts to define how best to use the facilities provided by the implementation environment. By its very nature Physical Design, if it is to be effective, demands a detailed knowledge of how to use the implementation environment. However, this has led, on occasions, to a rather undisciplined approach to implementation, falsely justified by the importance of product experience. The adoption of a Physical Design Strategy is an attempt to formalise and codify local experience and to integrate it more rigorously into the development effort.

PDS high level decisions may include:

- Whether to produce a generalised physical data design before tailoring it to the specific environment. If the application is likely to be implemented in several different target environments or in different locations, a generalised design will prove to be a good starting point for each individual implementation.

- Whether to produce an optimised physical data design directly from the logical design. If the design team is particularly familiar with the DBMS being used and the application will only be implemented for this environment, an intermediate generalised design will be less useful.

- How to calculate size and performance estimates. Estimation is difficult in that it requires detailed product knowledge, some of which may only be gained by experience. Reliable estimates are critical as errors can result in implemented systems which will not physically run on the target environment. Appropriate procedures must be well documented so that they can be applied uniformly and without wasted effort.

At a more detailed level other issues must be considered:

- The relative importance of such criteria as speed, usability, security, and maintainabilty must be determined. Decisions in this area are more likely to be relevant to a particular application rather than be a part of a more general strategy, although the development of local standards is clearly advantageous.

- The facilities of the target DBMS which may be used should be identified. If the application is to be ported over different software and hardware platforms, the design should be limited to the subset of facilities common to all of these.

- The standards and requirements for documentation and for the physical design procedures should be defined. Some development environments may provide automatic documentation facilities which, if appropriate, should be used to the full. Increasingly extensive data dictionary facilities are incorporated into proprietary DBMS's. These may provide an excellent capability for maintaining documentation concerning the data formats and structures and the associated integrity constraints.

10.3   THE LINKS BETWEEN PHYSICAL PROCESS AND DATA DESIGN

Development methods increasingly reflect the inter-relationship between process and data design. The processes have to manipulate the data as efficiently as demanded by the application requirements. This inevitably means that data design must proceed with one eye on process design and vice versa. This overlap is further blurred by current implementation strategies made possible by sophisticated software like modern Relational Database Management Systems (RDBMS). To some extent, such systems can not only manage the data but also the associated processing. Processes may only operate upon the data within certain business rules or integrity constraints, which are determined by the application and are identified during logical data modelling and entity-event modelling.

These integrity constraints may be implemented in two ways. The first is to implement them directly in the procedures, possibly using common routines for constraints which need to be enforced in several different circumstances. Thus the implementation of integrity constraints is distributed amongst the procedures.

This may have some undesirable consequences:

– it is difficult to verify that all the requirements have been implemented;

– modification to these business rules is a greater and more hazardous maintenance task.

Alternatively, the integrity constraints may be defined using the facilities of a RDBMS. Many RDBMSs provide automatic data vet at the type level. For instance, the RDBMS only permits the entry of numeric values if the entry field has been defined as numeric, or valid dates if the field has been defined as a date field. RDBMSs increasingly provide support for the enforcement of such general integrity constraints as Entity Integrity and Referential Integrity, either explicitly or utilising a combination of RDBMS features.

Entity integrity is enforced by not allowing a primary key to have null values. In the simplified patient database described in Figure 10.5, entity integrity could be enforced for the Patient entity (for example, only allowing the entry of a new patient when a unique patient identifier, Patient-number, is assigned). The consequences of enforcing entity integrity must be considered carefully. If a patient is admitted as an emergency through casualty, and if patients may not be allocated to a ward until the database has been updated, then the allocation of a unique identifier becomes a critical issue. If the allocation of the Patient-number is in any way protracted, the enforcement of entity integrity may affect the treatment of the patient.

Referential integrity states that the value of a foreign key must either wholly match the value of the primary key of one of the occurrences of the master entity or that the value of the foreign key must be null. Again, the consequences of imposing referential integrity must be thought through carefully. If a consultant leaves the hospital, referential integrity may be maintained by either deleting the records of all the admissions assigned to the departing consultant or making the Consultant-number data item null in all these admission records. The first strategy is not helpful as the patients are likely to still be treated by the hospital and the second strategy confuses these patients with those that have not yet been allocated to a consultant on admission. Of course, leaving Consultant-number unchanged may be equally confusing if data relating to the departed consultant is deleted. A simple solution to the problem would be to insert another attribute, Date-left, for the Consultant entity to identify those consultants no longer available to treat patients.

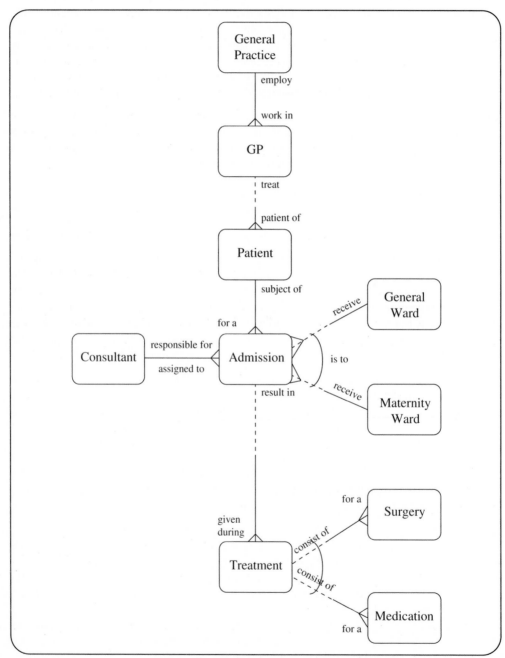

Figure 10.5   LDS: patient database

Typically, these general integrity constraints are defined declaratively in the RDBMS. This has the advantage that the constraint is defined once rather than each time a process has to operate upon the database.

Similarly, application specific integrity constraints may also be described in some database management systems. A constraint such as

'A borrower may not borrow further books if the borrower has an outstanding fine.'

may be specified once in the DBMS and is invoked automatically when a new loan is to be recorded for a borrower. The DBMS not only encompasses the data definition but also the definition of all the rules governing the manipulation of the data.

There are difficulties with this approach. First, the integrity constraints may be too complex to be expressed within the database language. Secondly, the automatic features of the DBMS may make it difficult to deal with exceptional cases – once the integrity constraint is enforced, it may be enforced without exception.

It is thus important to identify those circumstances when it is appropriate to utilise the more sophisticated features of the implementation environment. As the descriptive power of modern DBMSs increases and their performance continues to improve, they will be used to encompass more aspects of the implementation. Rules traditionally specified in programs will increasingly be found together with the data definitions.

10.4   CONFLICTING REQUIREMENTS OF PHYSICAL DATA DESIGN

Physical Data Design should, in theory, attempt to minimise the following factors:

- data access times;
- processing times;
- use of backing store;
- system development time;
- system maintenance time;
- the need for database reorganisation;
- system and user interface complexity.

These factors frequently represent conflicting requirements. For example, the use of backing store may be reduced but result in greater system complexity and perhaps increased maintenance times. SSADM suggests some general principles for assigning priorities to these conflicting design criteria. Trends in hardware and software costs suggest that the physical design should be based upon the following assumptions:

- backing store costs are less significant than development and maintenance costs;
- run times and response times should be minimised to satisfy user interface and operational requirements;

- development and maintenance costs should be minimised;
- the end-user interface should be kept as simple as possible, particularly when physical data is being manipulated directly by the user using advanced DBMS features.

The last of these objectives may be best achieved by maintaining a one-to-one mapping between the logical and physical designs. The first-cut physical data design, does this by being directly derived from the logical data design, which is a succinct and clear description of the data requirements.

When, for performance or other reasons, the physical database has a different structure from the logical database two alternative strategies exist:

- the logical database processes previously defined may be enhanced to handle the physical structure;
- the logical database processes are implemented as originally defined and a Process Data Interface (PDI) is introduced which maps the logical structure onto the physical structure and transfers the data from the physical database to the processes.

These issues are discussed further in Chapter 11.

10.5   LOGICAL/PHYSICAL DATA INDEPENDENCE

The Required System LDM represents an unambiguous and implementation-free description of the data. This is likely to be a more accessible and stable view for the user than the description of the actual physical implementation which may vary from time to time as the implementation platform changes. This separation of logical and physical perspectives should make maintenance easier. Changes which are only concerned with physical database access may be performed without any change to the logical processing. The converse is also true. This again reinforces the advantage of a PDI when the physical database is different from the logical database.

The SSADM universal function model shown in Figure 10.6 illustrates the relationship of the Process Data Interface with the other components of the system.

10.6   ACTIVITIES IN PHYSICAL DATA DESIGN

Physical Data Design is achieved by applying a series of operations to the Required System LDM. The first seven activities are based upon certain assumptions concerning the nature of database management systems in general. It should be appreciated that not all these assumptions are relevant to design for all types of DBMS. For instance, the DBMS may offer no facilities for controlling how records will be physically blocked.

The general assumptions underpinning Physical Data Design are:

- entity types correspond to record or table types and entity occurrences to record or row occurrences;
- the unit of physical access is the block or page;

- records which are frequently accessed together should be physically placed near each other to reduce access time;
- the primary relationships between different entities is well supported;
- the secondary relationships between different physical groups may be less well supported and necessitate additional design effort.

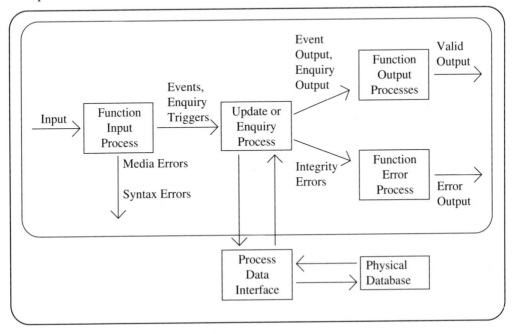

Figure 10.6 Universal function model with PDI

Relationships between entities in the same physical block are known as primary relationships, whilst those between entities in different physical blocks are secondary relationships.

The activities that comprise Physical Data Design are listed below:

- identify the features of the Required System LDM which are relevant to the physical data design;
- document the required entry points indicating which are non-key;
- identify the roots of physical groups;
- identify allowable physical groups;
- apply the least dependent occurrence rule;
- determine the block size to be used;
- split the groups to fit the chosen block size;
- apply product specific rules.

If certain activities can be identified as irrelevant because some of the assumptions are not applicable to the chosen DBMS, then they should be omitted from the prescribed physical data design procedure of the Physical Design Strategy.

The patient database shown in Figure 10.5 will be used to illustrate the activities.

A: Identify the features of the Required System LDM relevant to physical data design.

An easy way of documenting the requirements for physical design is to produce a physical version of the LDS. This is only a working document and does not replace the LDS. It serves to distinguish the physical design issues from the logical design and is simpler than the LDS.

The local requirements for the physical data structure documentation should be specified in the PDS. Typically, the differences between the LDS and the physical data model are:

- relationship names need not be recorded on the physical data structure;

- physical design is only concerned with one aspect of optionality, whether or not a master entity must exist for a detail entity. Mandatory detail to master relationships are used to determine the placing of non-root entities in physical groups;

- exclusivity arcs on relationships may not affect physical data design if its scope is limited to the placement and access of data. Exclusivity constraints may be implemented procedurally.

The design volumes are transferred from the LDS to the physical data design. These show the number of occurrences of each entity (shown in each entity box) and the number of details per master (shown near the crows' feet as in Figure 10.7).

The physical data structure for the patient database is shown in Figure 10.7. The entities are represented by hard boxes to distinguish physical from logical designs. Optional relationships at the detail end are denoted by open circles. For instance, the relationship between Patient and General-Practitioner is optional to reflect that a patient's GP may not be known when the Patient record is created (the patient could be unconscious).

B: Document the required entry points indicating which are non-key

Non-key entry points will be translated into secondary indexes in the implemented system and are recorded on the physical data model. They are identified by:

- deriving the access entry points from the Effect Correspondence Diagrams and the Enquiry Access Paths;

- indicating the required access points on the physical data model, denoted by a circle with an arrow pointing to the appropriate entity with the relevant data items listed;

- comparing the data items listed for each access with the key of the entity concerned to identify those which are non-key;

– indicating those that are non-key using a lozenge-shaped box linked to the entity it is indexing.

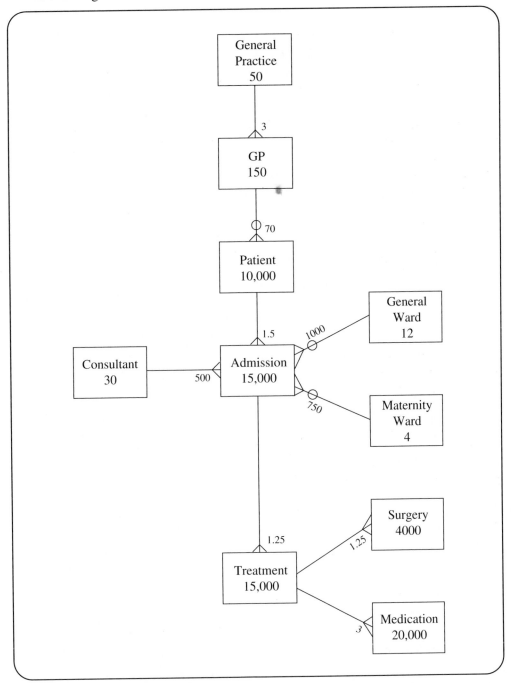

Figure 10.7 Physical data structure: patient database

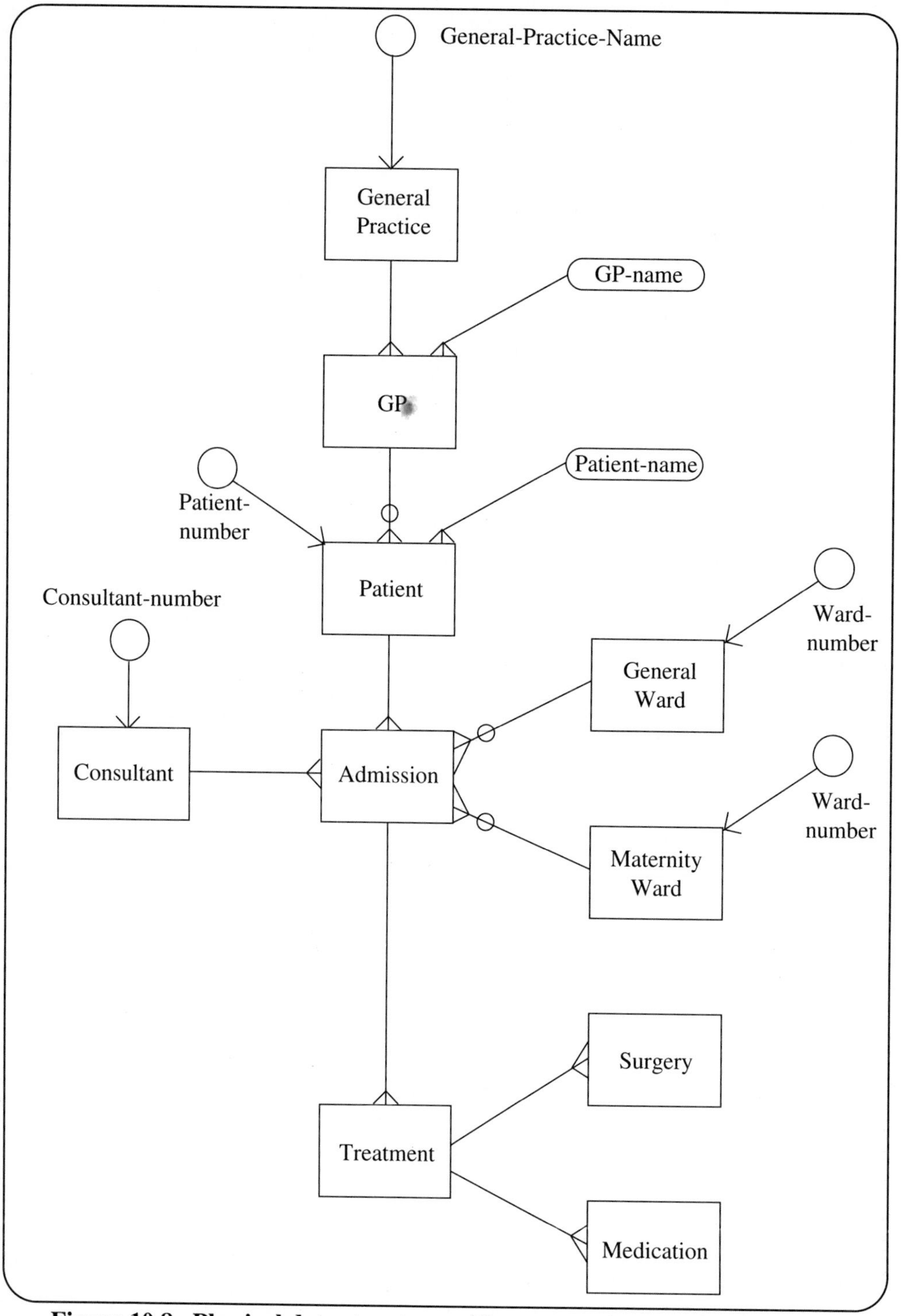

Figure 10.8   Physical data structure with access points: patient database

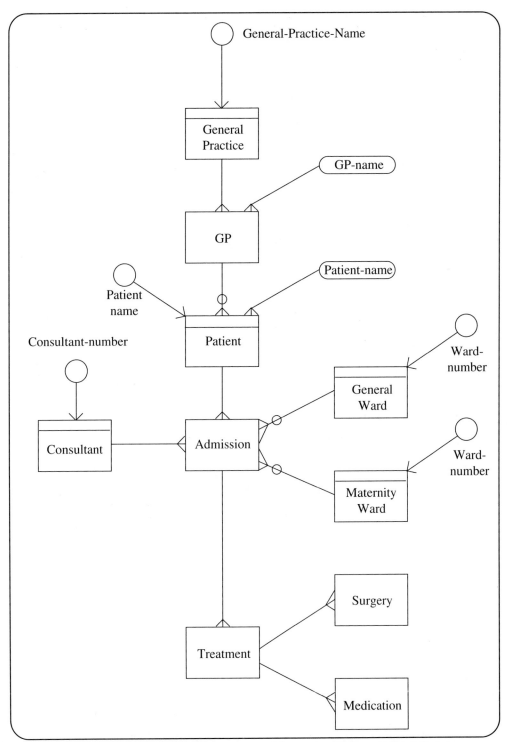

Figure 10.9 Physical data structure with root entities: patient database

In Figure 10.8 the access points have been added to the physical data structure. The non-key access by Patient-name is required to make retrieval possible when Patient-number is not known.

C: Identify the roots of physical groups

There are two kinds of root entity which may become the root of a physical group:

- entities which do not have any master entity in the data model; these are known as reference entities;

- entities used as direct access points except where the entity has an hierarchic or compound key and one of its master entities, whose key is part of the access key, is already top of its group.

Root entities are denoted by a stripe across the top of the entity box.

Root entities are identified on Figure 10.9. The entities General-Practice and Consultant are examples of reference entities and the Patient entity is used as a direct access point.

D: Identify allowable physical groups

The physical groups to which non-root entities are allowed to belong are determined by applying the following criteria:

- a non-root entity may only belong to a physical group which contains one of its mandatory masters;

- if a non-root entity is a direct access point it should be placed in the group which contains the entity whose key is part of its own.

The allowable physical groups are shown in Figure 10.10. Entities may satisfy the criteria to belong to more than one physical group. The entity Admission could belong to either the physical group with Consultant as its root or the physical group with Patient as its root.

E: Apply the least dependent occurrence rule

The least dependence occurrence rule is only a rule of thumb but produces acceptable results. It states that if an entity may be placed in more than one physical group (after applying the criteria in Step 4) then it should be placed in the group in which it occurs least. This helps when a place near facility is offered by the DBMS.

The application of the least dependent occurrence rule to the patient database suggests that the entity Admission is better placed in the physical group with Patient as its root (Figure 10.11).

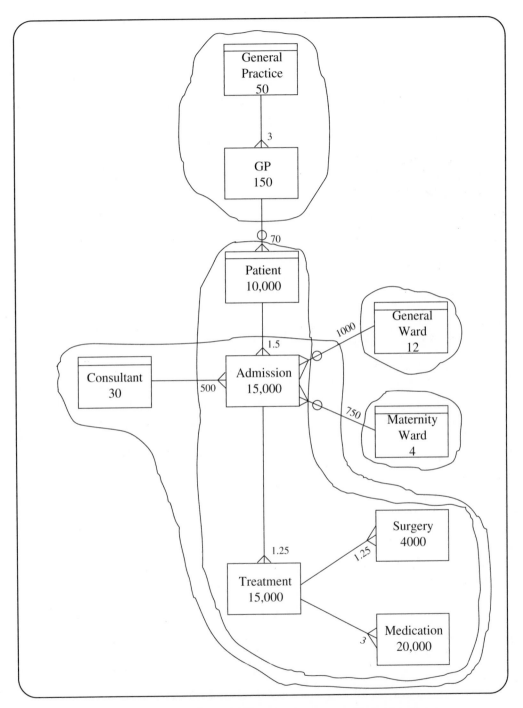

Figure 10.10   Physical data structure: patient database

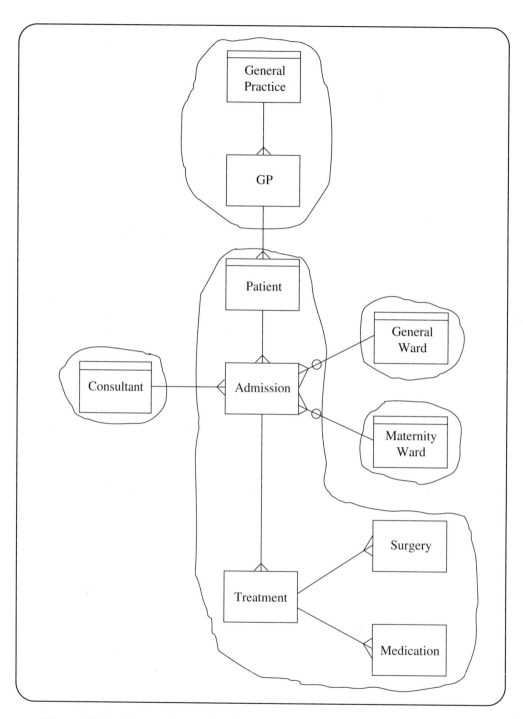

Figure 10.11   Physical data structure with physical groups: patient database

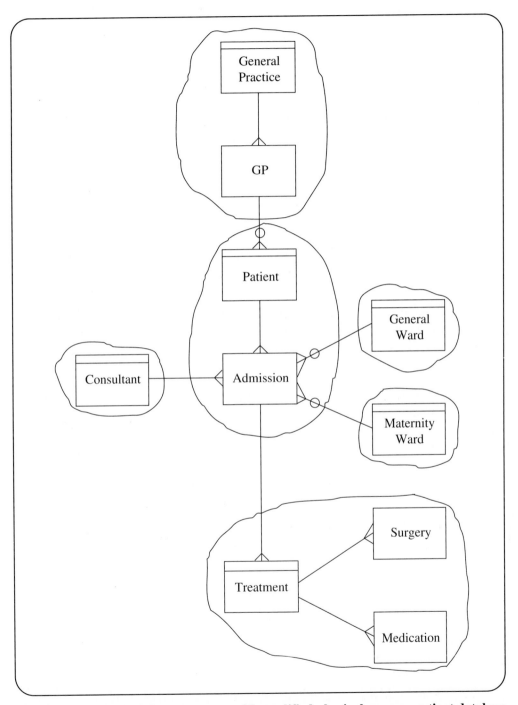

Figure 10.12 Physical data structure with modified physical groups: patient database

F: Determine the block size to be used

Several factors affect the choice of block size for first-cut Physical Data Design:

- the block sizes supported by the DBMS;
- the sizes of the most commonly used physical groups;
- the amount of space the blocks will take up when read into memory.

Ideally, the block size should be chosen to accommodate the largest of the commonly used groups. These may be identified from the Function Definitions and the associated ECDs and EAPs. Thus in the patient database example the Patient physical group should determine the block size. However, the block size should not be so large that the necessary buffer size causes memory problems when the system has to cope with a number of transactions concurrently.

G: Split the groups to fit the chosen block size

If a physical group cannot fit into the selected block size, the group must be split. This is best achieved by working up from the bottom of the existing group to identify a suitable sub-group which will fit into a disk block.

For the patient database it might transpire that the Patient physical group is too large. This problem could be resolved by identifying a new physical group with Treatment as its root (Figure 10.12).

H: Apply product specific rules

Preliminary decisions about which DBMS facilities should be used may already be documented in the Physical Design Strategy. These should be reconsidered in the light of further information gleaned during the application of the previous steps. When the product specific rules have been defined, they are applied to produce a product specific data design.

For example, for a RDBMS the following principles may be applied:

- each record type in the first-cut physical data structure becomes a base table. Each record type is expected to be in third normal form;
- each base table is indexed by its primary key;
- each compound-key base table is indexed by each element of its compound key;
- indexes are defined for each foreign key and each non-key access requirement.

Mechanisms for 'place near' support vary from one RDBMS product to another. If a clustering index is supported, which ensures that the data is physically stored in the sequence of the index, then:

- for each non-root base table, the foreign key from its owner in the first-cut design hierarchy is defined as a clustering index.

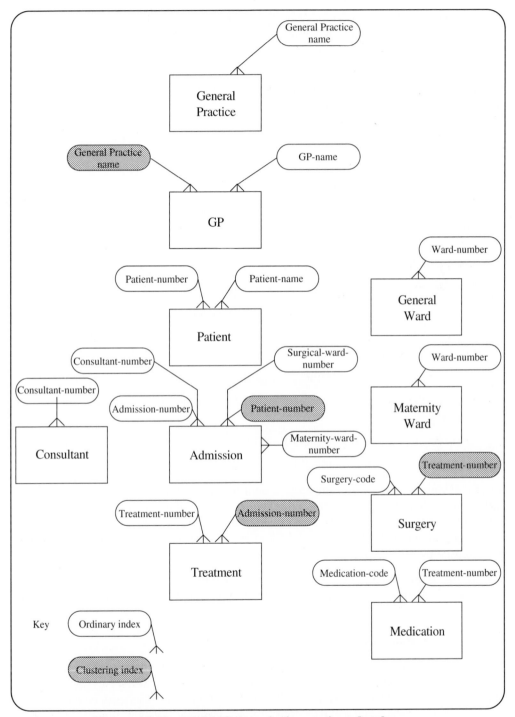

Figure 10.13   RDBMS data design patient database

This will ensure that the record type occurrences with duplicate values of the index key are located in the same disk area. Other place near mechanisms may be concerned with record types (tables) sharing the same tablespace or physical unit of storage (block or page). Care should be taken to use the vendor's product specific rules when determining the actual Physical Design Strategy.

The RDBMS data design for the patient database is shown in Figure 10.13. In this example Patient-number is defined as clustering index for Admission as it is the primary key of Admission's owner in the hierarchy.

10.7   ESTIMATING SYSTEM SIZE AND PERFORMANCE

Size estimation is of critical importance. The size estimation forms previously prepared during the Physical Design Strategy are now used. The likely number of instances of each entity type has been estimated earlier. The size of the application system may be estimated by the following steps:

- determine the size of each data item in each record or table type;
- calculate the size of each record or table type;
- determine the size of each index (both primary and secondary) for each record or table type;
- calculate the storage requirement for all the record instances for each record type;
- calculate the storage requirement for all the index instances for each record type;
- calculate the storage requirement for all data by summing the requirement for each record type;
- calculate the storage requirement for all the indexes similarly;
- the total storage requirement is the sum of the data and index requirements.

It is not uncommon for the storage requirement for the indexes to exceed that of the data, particularly when the application is indexed by a large number of character fields. To allow for load factors and other operational requirements, such as security, log files, etc, it is suggested in the SSADM Version 4 Manual that the total storage requirement should be increased by 60%.

The storage requirements for the developed application software and the target system software should also be included when estimating the total storage requirements.

10.7.1   Estimating the Access Times of the Major Functions

The first-cut physical data design must be tested against the processes to determine whether the application service level requirements are met. Satisfying performance objectives is a critical factor when the success of the application comes to be judged. This is true for both batch oriented and on-line systems. For batch systems, unreasonably long processing times may make the system inoperable or place too much demand upon the computing facility. For on-line interactive systems, a good response time is essential if the system is to maintain the confidence of the users. Thus it is important to estimate the system performance so that any problems may be identified before the system is actually constructed.

Frequently a form of Pareto principle applies to computer systems, in that the majority of the activity of the system is performed by a small percentage of the transactions in the system. It is also likely to be the case that only a small number of the transactions need to meet time-critical deadlines. Typically approximately 20% of the transactions should be examined for the purpose of performance estimation. The critical transactions can be identified quite easily. For instance, those transactions that create or update the entities with large numbers of occurrences, commonly those detail entities towards the bottom of the LDM, are likely to be critical. When the transactions have been identified, each must be examined in turn.

For the purpose of the first-cut estimates, it is assumed that the accesses across the physical data structure accurately reflect the operation of the transaction. A logical read operation will be assumed to require one physical read operation (each of which may generate several disc assesses). Direct access to a record which is not the root of a physical group may require several read operations before the required record is found.

For each transaction, the timing estimation form should be completed. This should show where possible:

- the physical access information;

- the disk accesses to perform each read and write;

- indications of CPU time used (this may be difficult to gather, although in critical situations prototyping can be very useful).

10.8   SATISFYING PERFORMANCE CRITERIA

The performance of the application system is clearly important, but it must be assessed in the context of the documented user requirements and not in terms of some abstract view of ideal performance. In order to emphasise this, SSADM suggests only two guiding principles for optimisation:

- optimise only to achieve agreed requirements;

- minimise the differences between the LDM and the Physical Data Model.

10.8.1   Optimise Only to Achieve Agreed Requirements

If the application system meets the user's performance requirements, unnecessary optimisation may have unsatisfactory long term side-effects. Optimisation, by its very nature, may destroy the one-to-one correspondence between the logical model, and the physical implementation. Unlike the logical model, the physical design may have to be biased to optimise the performance of one set of transactions at the expense of another set. This almost inevitably makes the application system less flexible and maintainable.

10.8.2   Preserve the Logical Structure

The Logical Model of the system, embodied in the Required System LDM, has been developed so that it is:

- clear and unambiguous;

- free from unnecessary redundancy;

- easy to enhance and extend.

Nonetheless, it may be impossible to implement the logical model without modification because of:

- performance constraints;

    or

- limitations in the target software and hardware.

If end-user query facilities (Query-by-Example:QBE or Query-by-Forms:QBF) are to be used, then the database structures may be visible to the user and hence should be kept as simple as possible. This places a significant limitation upon the degree of optimisation possible. If the DBMS permits the user to see a logical view of the system which is different from the actual implementation, without critical performance penalties, the problem is more tractable. For instance, many RDBMSs permit the definition of views which may be derived from several base tables.

Where possible, optimisation should utilise the features of the DBMS to satisfy agreed criteria and only as a last resort disturb the one-to-one correspondence between physical and logical models.

10.9   OPTIMISATION WITH DBMS FEATURES

Optimisation criteria may be conflicting. For instance:

- reducing the required data storage (using a data compression technique, for example) can increase processing time;

- minimising process time may necessitate an increase in the data storage requirement (eg introducing derived data);

- minimising space or processing time can increase the system complexity (eg unnormalising the relations).

Thus optimisation involves making a series of compromises with different performance criteria and different sets of transactions.

Storage requirements may be minimised by using such features as:

- variable length data items;

- space compression in textual fields;

- variable length records;

– adjustment of block size and access methods.

Equally performance criteria may be achieved by optimising access to data storage by:

– placing details near their master entity occurrences;

– appropriate selection of access method;

– implemention of key only entities as indexes;

– providing secondary indexes for foreign keys;

– providing indexes for non-key data items used for access;

– building required indexes at run-time for a process;

– providing a direct access mechanism to detail occurrences;

– holding small reference tables in main memory during process execution;

– adjusting block size;

– adjusting number of blocks in buffers;

– adjusting blocking factor.

It is also important to check that when the DBMS is optimised with respect to one process, the speed of execution for other processes has not been inadvertently reduced to an unacceptable level.

10.10 OPTIMISATION BY TUNING THE PHYSICAL MODEL

If the implementation facilities of the DBMS do not provide sufficient performance, it will be necessary to compromise the structure of the physical model so that the implementation is no longer directly based upon the LDM and the Logical Database Processes. Tuning the physical model to satisfy specific performance targets may produce a system which is more difficult to maintain and less easily supports ad hoc user queries. Optimisation may compromise the logical design in two ways, by modifying the database structure or the Logical Database Processes.

The first-cut physical data structure may be modified to enhance the performance of a particular transaction by:

– introducing redundant or derived data in the form of counts or totals, or duplicating data in more than one table to reduce the access path or the number of accesses for a particular transaction;

– combining logically separate entities to reduce the number of database accesses for the transaction;

– splitting a single entity so that a smaller volume of data is transferred during the transaction;

– adding relationships to reduce the number of entities that have to be traversed during the transaction.

10.10.1   Introducing Redundant Data

In Figure 10.14, the physical data design fragment shows that a salesperson is assigned to many customers. One of the transactions may be required to output the number of customers assigned to, and the total customer-debt for, a particular salesperson. This can be supported by the physical data structure as it stands by counting all the customer records for a salesperson. If there are a large number of customers for a salesperson, then the process execution time may be unacceptably long. This can be significantly reduced if additional data items, Number-of-Customers and Total-Debt, are introduced into the salesperson record. Unfortunately, this change produces a performance reduction for other transactions. For example, the salesperson record must also be updated whenever a new customer is added, an existing customer deleted or a customer debt changed.

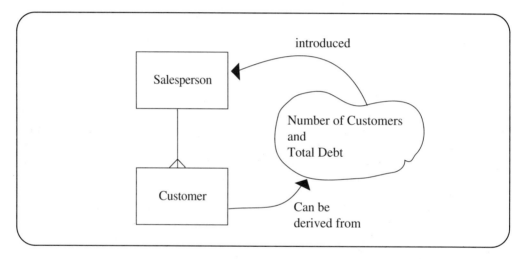

Figure 10.14   Introducing redundant data

10.10.2   Combining Logically Separate Entities

Figure 10.15 shows a physical data design fragment for a system which records the employees assigned to each office. A transaction is required to display the names of the occupants of a specified office. The transaction is supported by the physical data structure but requires access to the office record and several employee records. If it can be assumed that no more than four employees will ever be assigned to an office at any one time, then the speed of the transaction may be improved by combining the entities. Additional attributes are defined in the Office entity to store details for a maximum of four occupants. However, this solution has disadvantages. The records for offices which have not been allocated four employees will contain null values, resulting in a possible waste of space. Similarly, a change in company policy to permit five employees per office cannot be supported by this data structure.

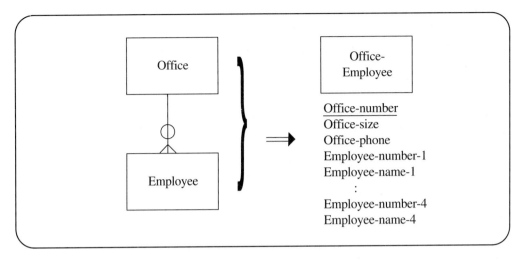

Figure 10.15 Combining separate entities

10.10.3   Splitting Entities to Reduce Data Transfer

Figure 10.16 gives a physical data design fragment of a system which records data concerning vehicles owned by a transportation company, and at which depot they are maintained. Two transactions have been identified during logical design. The first only requires details concerning the vehicles current location and loaded weight, whilst the second requires the servicing information for the vehicle. If the transactions have to operate upon a database where all the vehicle data is held in one entity, then both transactions will involve the transfer of unwanted data. This difficulty is eliminated if two entities are defined – Vehicle-Use for the first transaction and Vehicle-Service for the second. However, if a transaction requires all the vehicle data it must access three entities. The modified physical data design is shown in Figure 10.17.

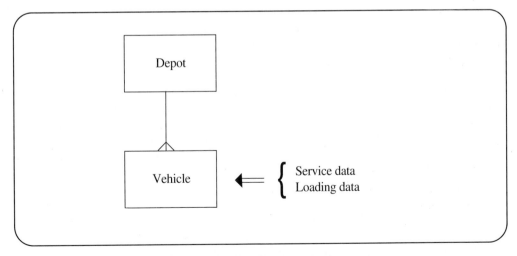

Figure 10.16   Splitting entities

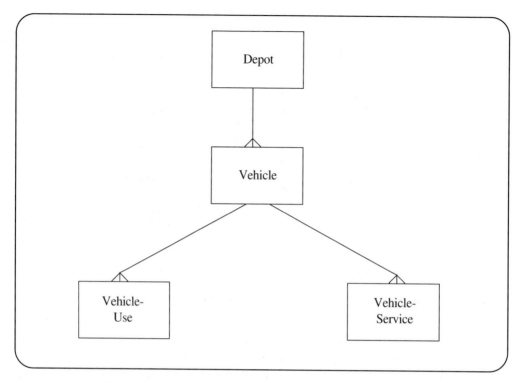

Figure 10.17   Splitting entities – modified design

10.10.4   Adding Relationships

In a simple sales-order processing system, sales staff are assigned to customers and customers place orders. The data design fragment is shown in Figure 10.18. An order enquiry needs to display the name of the salesperson concerned. Currently, the model requires access to the Salesperson entity via the Customer entity. The access path for the transaction is shortened by adding a relationship directly between the Salesperson and Order entities. Figure 10.19 shows the modified physical data design. Again, this increases the execution time for the transaction which creates the orders, as referential integrity checks are needed against the Salesperson entity.

Inevitably these changes to the data structure will require the Logical Database Processes to be modified or the introduction of a Physical Database Interface to present the processes with a logical view of the database. Some optimisation may also be achieved by changing the operation of the processes by:

- postponing database updates and storing them in transaction files for later off-line processing;

- postponing database deletes, again for later off-line processing;

- organising processing to minimise the number of database accesses by a careful choice of access path.

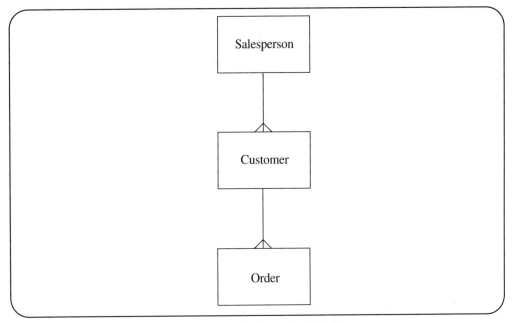

Figure 10.18   Adding relationships

Once the database structure has been optimised, it is important to examine the processes again to ensure that the optimisation procedure has not had a disastrous effect upon their execution. Optimisation is concerned with making the most appropriate series of compromises for the requirements of the application.

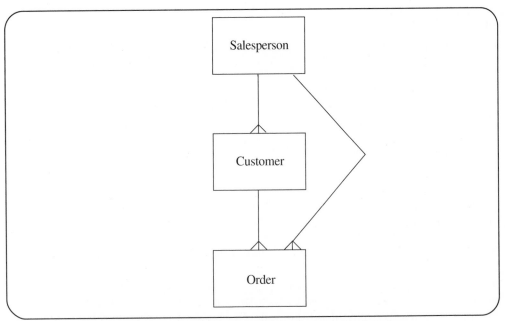

Figure 10.19   Adding relationships – modified design

The user must be informed if it is not possible to achieve the required performance using the selected target hardware and software. This may result in:

- modification in the overall system requirements so that the necessary performance may be achieved;
- a lowering of the performance requirements;
- improving the performance of the implementation hardware;
- improving the performance of the implementation software.

10.11   DOCUMENTATION

Physical Data Design is documented in several ways. The elements of the Physical Environment Classification relevant to data design are completed. The Function Definitions are updated with any sorting requirements to optimise performance. Any processing optimisation requirements that have been identified are recorded in the Requirements Catalogue. Changes to the service level requirements are documented in the Function Definitions or the Requirements Catalogue. Timing and space estimation forms are also completed. The Physical Data Design is the final product and is documented according to the Application Development Standards. It is useful to record any major or complicated optimisation decisions because, when the system is maintained, it is important to know why certain design decisions have been made. SSADM provides no formal documentation standard for recording design decisions and this should be addressed when defining the Application Development Standards.

10.12   SUMMARY

Physical Data Design translates the Logical Data Model into a physical data structure for a specific implementation environment. This physical data structure must effectively and efficiently support the required system functionality. The significant variation in types of implementation environment requires an approach which may be customised for a particular environment. SSADM Version 4 encompasses this variation in approach by requiring the developer to define an environment specific Physical Design Strategy. This is developed after the Physical Environment Classification has been completed.

Physical Design proceeds initially producing a first-cut physical data model. A design based upon general principles is produced first and then product specific rules, documented in the PDS, are applied. This design could be implemented but may not satisfy the performance requirements for the system. These are achieved by first optimising the design utilising the DBMS tuning facilities. If satisfactory performance is not achieved by this means, then the physical data structure is modified to provide the necessary performance for the critical transactions. The physical data structure should not be modified unless absolutely necessary, as the one-to-one correspondence between the logical and physical data models is destroyed, resulting in a more complex implementation. Optimisation involves making design trade-offs between conflicting requirements, and is iterative. When the data structure has been optimised, it should be tested against the transactions to ensure that they all provide satisfactory performance. Finally, simulation can be used to validate the performance of the final design.

Physical Data Design utilises many of the products of logical design. The relationship of Physical Data Design to other SSADM models is shown in Figure 10.20.

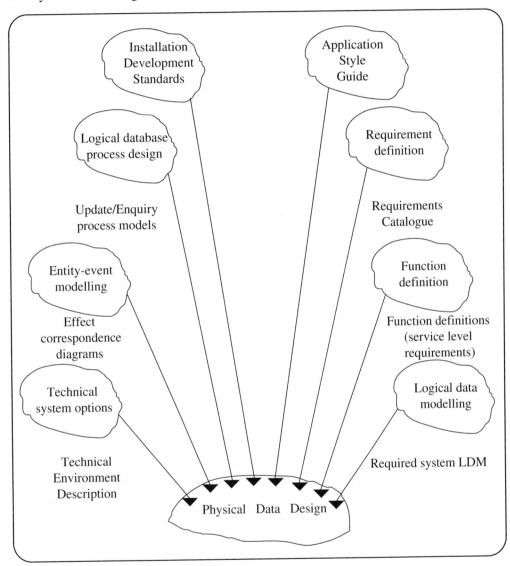

Figure 10.20 Physical data design and other models

Physical Data Design in Version 4 differs from Version 3 in that it explicitly provides a pragmatic practical design strategy for differing implementation environments. The design process need contain no irrelevant procedures, as it can be tailored in a methodical manner to suit particular implementation environments.

11 Physical Process Specification

11.1 INTRODUCTION

Physical process specification is concerned with the conversion of the models produced during logical design into program specifications, physical I/O formats and physical dialogues. The Enquiry Process Models (EPM) and the Update Process Models (UPM) produced by Logical Database Process Design are translated into program specifications. The I/O Structures form the basis of physical I/O formats, and the Logical Dialogue Designs are used for Physical Dialogue Designs for the target hardware and software. Figure 11.1 illustrates these relationships.

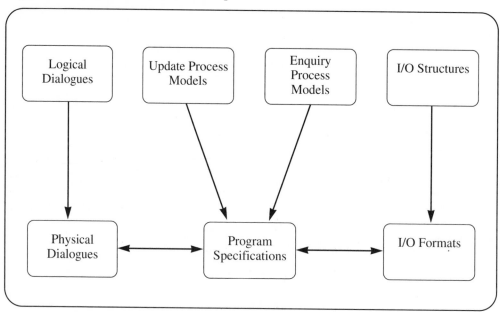

Figure 11.1   Development of physical process specification

Physical Process Specification, like Physical Data Design, is not a tightly prescribed activity but general guidelines are provided by SSADM. These guidelines have to be customised to suit particular combinations of:

- development hardware and software;
- target hardware and software;
- project constraints.

These guidelines are encompassed in the Physical Design Strategy (PDS).

The prime objectives of Physical Process Specification in SSADM are to :

- complete the parts of the Physical Design Strategy relevant to process specification;
- complete the specification of the functions derived in logical design;
- specify the required processing to a level necessary for implementation.

These objectives are achieved by iteratively performing the processes of:

- design;
- testing against objectives;
- optimisation;
- review.

Physical Process Specification involves four main activities:

- preparing for Physical Process Specification;
- creating the Function Component Implementation Map;
- specifying the functions which are to be implemented procedurally;
- developing the Process Data Interface.

11.2   PREPARING FOR PHYSICAL PROCESS SPECIFICATION

This complements the definition of the remaining elements of the Application Development Standards and, as with preparing for Physical Data Design, may only require major effort if either the application type or the implementation environment is new in some respect. There are four activities associated with preparing for Physical Process Specification:

- investigating the implementation environment;
- preparing the Processing System Classification;
- specifying the Application Naming Standards;
- completing the Physical Design Strategy (PDS).

11.2.1   Investigating the Implementation Environment

The need to be fully aware of the facilities offered by the development and operational environments is obvious. Nevertheless, because of the pressures of deadlines, the cost of training and the availability of development staff, the detailed design of projects is frequently attempted with a less than adequate knowledge of the software tools being used. It is sometimes hoped that the necessary expertise will be learnt during development. Certainly, it is true that when a project using an unfamiliar implementation environment is completed the development team may well have developed the necessary skills. However, there is no guarantee that they will have been utilised effectively on the completed project. It is not uncommon for developers to realise how much simpler the development would have been if they had better appreciated the power of facilities offered by the implementation environment. The SSADM Version 4 approach to Physical Design is pragmatic and emphasises the need to understand the implementation environment prior to system development.

11.2.2   Processing System Classification

The Processing System Classification provides a uniform approach to capturing this environment specific knowledge so that it may be used effectively by the development team as well as supporting later maintenance. This information may be supplied in a usable format by the product vendor. If not, it should be documented on suitable forms and a typical document is shown in Figure 11.2. The Procedural\Non-Procedural box indicates that processes may contain both procedural and non-procedural code. The ticked Enquiry and Update boxes record that both update and enquiry processes may be generated. Detailed guidance upon how to complete this form can be found in the SSADM Version 4 Manuals.

Processing System Classification					SSADM Version 4
Project/System	Author	Date	Version	Status	Page of

Classes of tool feature **RDBMS with 3GL embedded SQL access**

Procedural/non-procedural [✓]   On-line/off-line [ ]

Success units

automatic commit or explicit commit under programme control

Error handling

default data type error handling

Process components

process components linked before run-time

Database processing

Update [✓]   Enquiry [✓]

I/O processing

multiple windows

Dialogue processing

default form generation

Dialogue navigation

hierarchical menu structures; direct links between dialogue

Process data interface

views supported

Distributed systems

SSADMv4 34

Figure 11.2   Processing System Classification

11.2.3 Application Naming Standards

Many software development environments use agreed naming conventions for the deliverables associated with physical design and construction. A good naming convention will make it easier to trace the development of particular elements of the system and identify those that will be used together. For example, a possible naming convention could be:

- System Type;
- System Name;
- Product Type;
- Product Name;
- Version Number.

The use of a naming convention can present artificial difficulties when categorising a system or product by type. Whether a product is of one type or another is not necessarily of importance. One processing routine designated as a library routine but which is only used once need not be a problem. However, if all processing routines are blithely designated as library routines irrespective of their use, the naming convention ceases to be of value. A pragmatic egoless approach is required, particularly as the agreed convention may be limited by constraints dictated by the physical environment. There is no standard convention suggested by SSADM for these Application Naming Standards and conventions appropriate to particular implementation environments should be developed by each SSADM user organisation. A catalogue of the documentation and products of the physical design should also be maintained. This catalogue provides a useful indexing facility which can help to identify potential re-use and aids maintenance. A typical entry in the catalogue might comprise:

- identifier of fragment/product/deliverable;
- class of fragment/product/deliverable;
- purpose/action;
- inputs/subject/pre-condition;
- outputs/object/post-condition.

11.2.4 Completing the Physical Design Strategy

The data design elements of the PDS have already been defined. Those that relate to process specification are completed by:

- determining the criteria for the selection of implementation routes;
- defining how these criteria should be applied to each product;
- customising the Activity Descriptions for Physical Process Specification to accommodate implementation environment specific features;
- specifying the Function Component Implementation Map (FCIM) standards.

The first two activities are concerned with clearly defining the factors which should be considered when proceeding with physical design. For example, it might be agreed that the implemented system should be portable across a range of software and hardware platforms. This factor, when applied to a particular development system, might result in a constraint upon the features of this system which will be used to guide implementation. Another example might be whether a procedural or a non-procedural implementation should be chosen for a particular type of function component. For example, the agreed local first cut design standard might be to non-procedurally implement all enquiry processes. If an RDBMS is being used this agreed standard would probably result in the use of SQL for enquiry processes.

The SSADM Activity Descriptions, which define the design procedures, are now customised to accurately reflect both the features of the particular implementation environment being used and the nature of the project. This is again an acknowledgement by SSADM that development procedures have to be tailored within a generalised framework.

The Function Component Implementation Map (FCIM) is a set of documentation which classifies and specifies the processing for the functions that the application will support. The FCIM is documented to local standards which have to be agreed. These standards may include:

- input/output programs;

- dialogue control programs;

- database programs;

- common modules.

Illustrative standards are given below.

Input/Output Program Specification

Program Overview	Scope and purpose of program.
Inputs and Outputs	The format, structure, data items, syntactic and control errors for input.
Fragments Used	Modules used with input and output data.
Procedure	Defined by applying JSP to the input and output data.

Dialogue Control Program Specification

Program Overview	Scope and purpose of program.
Inputs and Outputs	The format, structure, data items, syntactic and control errors for input.
Physical data groups/screen	A separate module may be developed for each of these with input and output data defined.
Procedure	The navigation between data groups may be defined using control tables.

Database Program Specification

Program Overview	Scope and purpose of program.
I/O messages	Data input to trigger the event enquiry, output and integrity error mesages.
Common Fragments	For example, PDI fragments with input and output defined.
Procedure	Supplied by the Enquiry and Update Process Models.

Common Module Specification

Module Overview	Scope and purpose of program.
Inputs and Outputs	Input data, success reply and error replies.
Procedure	Defined by applying JSP to the input and output data.

11.3 DEVELOPING THE FUNCTION COMPONENT IMPLEMENTATION MAP

When creating the FCIM the designer needs to work at two different levels:

– identifying duplication and commonality across all functions;

– adding further detail to functions by fully specifying the FCIM elements.

11.3.1 Removing Duplication

The processing of an event may, on occasions, be preceded by an enquiry which provides the user with information necessary to progress the event. Such an enquiry event pair offers the possibility for the elimination of duplicated processing by merging the two processes. Hence, the enquiry and update processes can be linked providing the user with one integrated interactive procedure giving:

– visual confirmation of the updates based upon the sub-enquiries;

– immediate error detection and reporting to the user.

This avoids the frustration of entering a 'screenful' of data before discovering that the first field entered was incorrect and this has invalidated the subsequent entries. Dialogue structures for this issue are discussed later in the chapter.

11.3.2 Identifying Common Processing

The search for common processing means that the specification process is iterative; completing the specification of a function, identifying common processing and as a result modifying the specification of other functions. There are no clearly defined criteria for identifying common processing, though elements of common processing are frequently one of the following types:

- conversion routines (eg imperial to metric units);

- validation routines (eg date validation);

- calculation routines;

- data access routines;

- formatting routines.

Indeed, the identification of common processing is pursued throughout the life cycle. The first opportunity occurs when the Elementary Process Descriptions are documented. As analysis and design proceeds the event and function level commonality can be distinguished and documented in Logical Database Process Design and Function Definition. Lower-level commonality, at the routine or procedure level, may be identified during physical design. Figure 11.3 shows the layers of the system at which common processing may be identified. Notice that a function may, itself, be used by several different super-functions. As was mentioned in earlier chapters, the relationship between functions and logical database processes is many to many. This diagram also serves to illustrate the role of the Process Data Interface which, if designed as a separate subsystem, comprises the database access processes at the lowest level in this hierarchy. The scope of this PDI is defined in the Physical Design Strategy.

The identification of common processing should be treated pragmatically. Care needs to be taken to ensure that common processing is not artificially and unproductively identified. Advanced DBMS offer facilities for the non-procedural definition of processing which is associated with particular data items. At the simplest level, data validation processing can be specified though calculation routines and integrity checking may also be supported.

11.3.3   Specifying the FCIM Elements

The complete specification of the FCIM elements is achieved by specifying:

- the physical success units;

- the error handling for syntax errors;

- the system controls and associated errors;

- the physical I/O formats;

- the physical dialogues.

A: Physical Success Units

Previously defined logical success units may not be appropriate for physical implementation because:

- they impose a performance constraint;

- they may result in locking the database for too long preventing other transactions

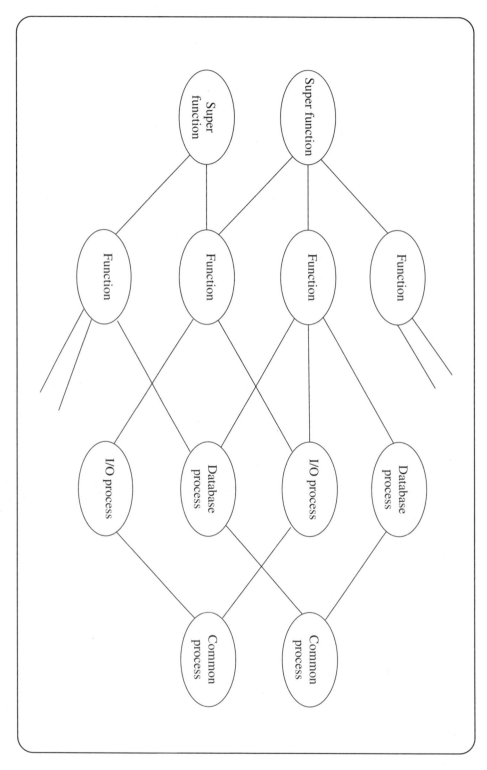

Figure 11.3 Layers of application system

from running;

- they represent too large a commit unit for the DBMS being used.

Each function should be examined and the appropriate level of the success unit determined given any constraints that may be imposed by the implementation environment. The choice of the physical success units may affect the operational characteristics of the processes and so user involvement is important.

For example, an application system to support tele-sales, recording customer orders on-line, could be unacceptably affected by an inappropriate record locking strategy. Such a system must respond rapidly to operator input and should also consistently display the most up-to-date stock details. If the complete order entry procedure for an order is one physical success unit, then the records for the products being ordered must be locked so that the stock quantity can be updated. This means that these records can not be accessed by other users whilst the order is being entered. Moreover, there is likely to be contention for the records of frequently ordered products even if a relatively small number of products are ordered on each order. However, if the entry of each item on the order is treated as a separate success unit, then record contention is less likely. Bear in mind, though, that when the details of the products ordered are modified before the order has been finalised, additional database accesses will be necessary.

B: Syntax Error Handling

Specifying syntax error handling at this stage rather than earlier in the design process has two advantages:

- the earlier design documents are less cluttered and focus upon the logical design;
- the method of handling these errors is frequently prescribed by the implementation environment.

The target system (eg DBMS) may perform some checks automatically and have in-built error handling and reporting features. For instance, it may be possible to include the error condition and error message in the data dictionary of the DBMS being used. Figure 11.4 gives an example.

data item	:	employee salary
error condition	:	not in the range 0.00 to 25,000.00
error message	:	invalid salary

Figure 11.4   Fragment Sample Data Dictionary Entry

C: System Controls

System controls fall into two categories:

- control of system navigation and execution;
- the control data errors.

The first category is concerned with ensuring that procedures are only executed in the correct sequence. This is a common requirement for batch processing systems, where the sequence in which programs are executed can be critical. For instance, in an accounting system the input of data for the new month may be prevented until the previous end of month reports have been produced. Similar constraints upon the sequence of processing exist for interactive systems. For example, a user may enter a sales order for a customer that may not yet exist in the database. In order to maintain referential integrity, a customer record should be created before the order record can be committed to the database. It is, however, awkward if the user has to abandon order entry to enter the missing customer record. It is much easier if a subsidiary update process can be invoked to handle the customer update without forcing the user to back out of order entry. In general, it is important that any operational constraints imposed by the system are consistent with the way the system can be most efficiently used.

Different user roles may have different levels of access to the system, requiring different paths through the menu structures. These controls on system navigation must be carefully planned to ensure that they support the activities performed by the user role.

Controls for the second category, data errors, may be implemented via:

- a dedicated audit function;
- the use of check totals during data entry and within the database itself.

An audit function may include transaction logging so that the effect of the transactions may be compared with the actual state of the database. Functions can be defined to read the data, making comparisons to ensure that the integrity of the database had been maintained. Transaction logging may be provided as an automatic feature of the target software being used.

Check totals are commonly used during data entry as batch totals which are calculated before entry and then automatically compared by the system after entry of the records in the batch. Equally, totals of values of the entity occurrences in the database can be maintained by the system and later subjected to verification by an audit program. Building this level of control and verification into the system incurs performance penalties. Whenever the database state is changed the totals also have to be modified, possibly doubling the number of database accesses. The impact of any performance reduction should be carefully balanced against the application requirement for check totals.

D: Physical I/O Formats

The definition of physical I/O formats relates to the production of output reports and input and output files (interactive dialogues are considered in the next section). The physical definition relies upon the earlier specification of the I/O Structure and the definition of the Application Style Guide. The preparation of physical I/O formats must involve consideration of ergonomic issues and the nature of the physical devices used such as the type of printer, the facilities offered by any file transfer system being used and the terminal type. Where the application is communicating directly with external computer systems, the relevant external standards (eg Banks Automated Clearing System or Electronic Data Interchange) must be applied.

E: Physical Dialogues

The physical dialogues for interactive processes are developed from the logical dialogue design and further specified to describe:

- physical screens;

- physical data groups;

- input views of a physical data group;

- dialogue components to handle errors;

- navigation paths through the dialogue.

The relevant project formats and procedures are defined in the Application Style Guide and should now be applied.

11.3.4   Physical Screens

Each physical screen may contain one or more data groups. Where possible the logical screens specified earlier should be implemented directly.

11.3.5   Physical Data Groups

Similarly, the physical data groups, which constitute the physical screen, may be direct implementations of the logical groups. However, this may not be possible for several reasons:

- the logical group is too large for one physical data group because of device limitations or constraints agreed in the Physical Design Strategy;

- the physical environment does not support the style of processing required by the logical data grouping, for instance windowing may not be supported;

- the logical group may be used for more than one input message each of which requires distinct processing.

Data may be grouped physically for several reasons:

- the data items are related to one entity;

- the data items are associated with one input message;

– the grouping simplifies the dialogue.

11.3.6   Input Views of a Physical Dialogue Group

Physical dialogues may be structured in three possible ways:

– one event is entered and then processed;

– a group of events is entered together and then processed individually;

– a group of events is entered and processed together.

These can be illustrated by considering a simple example.

The entry screen shown in Figure 11.5 corresponds with the physical data structure in Figure 11.6. Patient-number is entered and the process uses this to access the Patient entity and then search for the most recent admission. Patient-number is the primary key and Patient-name is displayed for visual confirmation. Consultant and Ward are foreign keys in the Admission entity and GP is a foreign key in the Patient entity. These foreign key data items may each be modified independently, corresponding to three different events:

– a patient is allocated to a different consultant;

– a patient is moved to a different ward;

– a patient is allocated to a different GP.

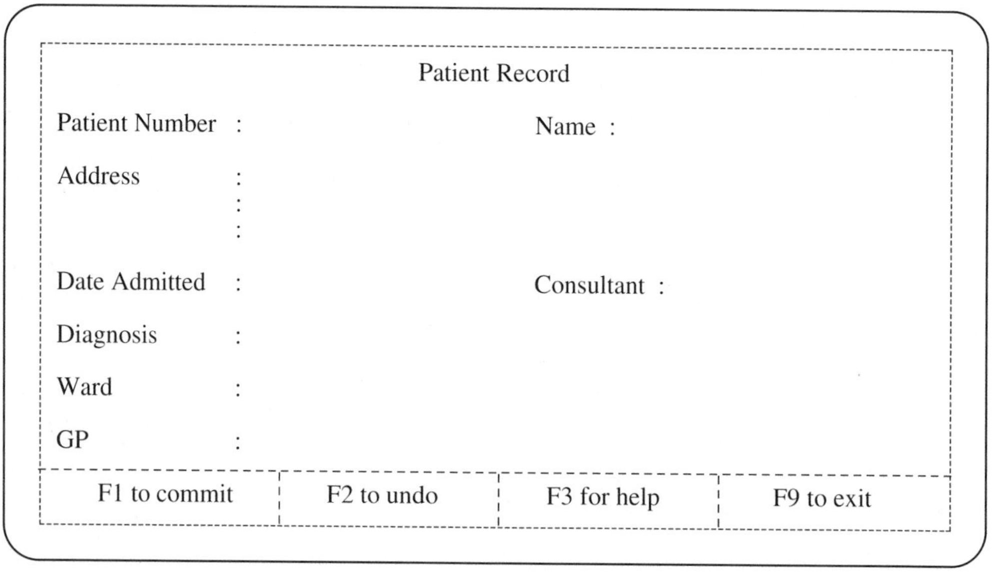

```
+---------------------------------------------------------------+
|                        Patient Record                         |
|                                                               |
| Patient Number  :                      Name  :                |
|                                                               |
| Address         :                                             |
|                 :                                             |
|                 :                                             |
|                                                               |
| Date Admitted   :                      Consultant  :          |
|                                                               |
| Diagnosis       :                                             |
|                                                               |
| Ward            :                                             |
|                                                               |
| GP              :                                             |
|---------------------------------------------------------------|
|  F1 to commit  |  F2 to undo  |  F3 for help  |  F9 to exit   |
+---------------------------------------------------------------+
```

Figure 11.5   Patient entry screen

The first strategy suggests that the user is prompted to select one of the foreign keys. The appropriate data group is presented so that only that event can be processed and, when completed, its success or failure is reported. For example, upon the selection of the foreign key Ward, a window may be displayed showing alternative wards, one of which may be selected. A value is chosen, is validated to check that there is an empty bed and if there is, the change is committed to the database.

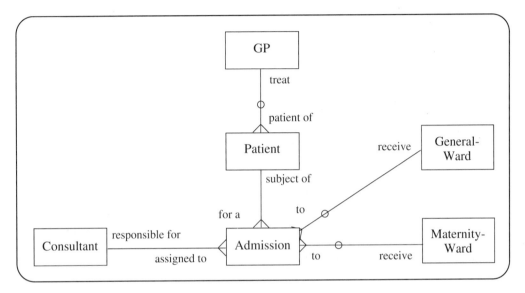

Figure 11.6   Physical data structure: patient database

The second strategy permits the user to modify one or more of the foreign keys. When the changes have been completed by the user, the dialogue detects which has been changed and invokes the appropriate Update Processes independently and in sequence. Messages are then displayed indicating the success or failure of each event. For example, if the values for Consultant, Ward and GP are all modified each of the changes would be completed in turn. If the value of Ward is invalid, then this change will not be committed to the database whilst the other changes will.

In these circumstances, the screen processing must be carefully planned to ensure that the user appreciates that all the processing has not been completed successfully, and changes which have not been made to the database have been clearly indicated.

The third strategy is to let the user make changes to one or more of the foreign keys. These changes are then processed in sequence. The processing stops as soon as one fails and all the other changes are undone. Effectively, all the processing of all the events is treated as one success unit or transaction. In this instance, an invalid value for Ward would prevent the other changes being committed to the database.

The first of these strategies is the most natural in that the user gets a single fail or success message after each event has been processed. However, hardware and software constraints may force the designer to choose the second option. If the third option is the most appropriate, the designer should carefully examine the dialogue to see if there are complex integrity constraints which had not been previously identified. This may suggest that the dialogue should be restructured. If a consultant may only treat patients in certain wards a more helpful dialogue would, when a different consultant is assigned to a patient, automatically prompt the user to select one of the consultant's wards if the existing ward is invalid. Furthermore, it could warn the user if there were no available beds in any of the consultant's wards.

11.3.7   Error Handling

The dialogue must be able to cope with the errors that may occur. These have been discussed earlier in this Chapter and in Chapter 9. Typically, the dialogue may be required to handle three types of errors:

- syntactic errors;

- control errors;

- integrity errors.

Where possible, the implementation software (eg RDBMS) can be used to trap and report these errors. On occasions when an error condition is unique to a particular function, the error handling should be specified in the physical data input specification for that function. Some DBMSs provide features which allow general integrity constraints to be overridden in exceptional circumstances.

11.3.8   Navigation Paths through the Dialogue

This specifies how the user may navigate around a particular dialogue as well as the navigation between dialogues. The physical implementation environment should be utilised where appropriate, permitting the user flexible use of the application. When several on-line functions are always used in sequence they can be linked either directly or via proximity in the menu tree.

If data for the same entity or for a group of linked entities is to be entered, direct links will be easier for the user. As the user is taken from one dialogue to the next, data items are automatically carried forward to avoid re-entry. An order entry sub-system is a good example. The user may be presented with a screen to enter the order header details which, upon completion, is followed by a screen for the individual order-line details. Figures 11.7 and 11.8 show illustrative screen layouts. The Order-number, Order-date, Customer-number and Customer-name are automatically carried from the Order Header Entry screen to the Order Line Entry screen.

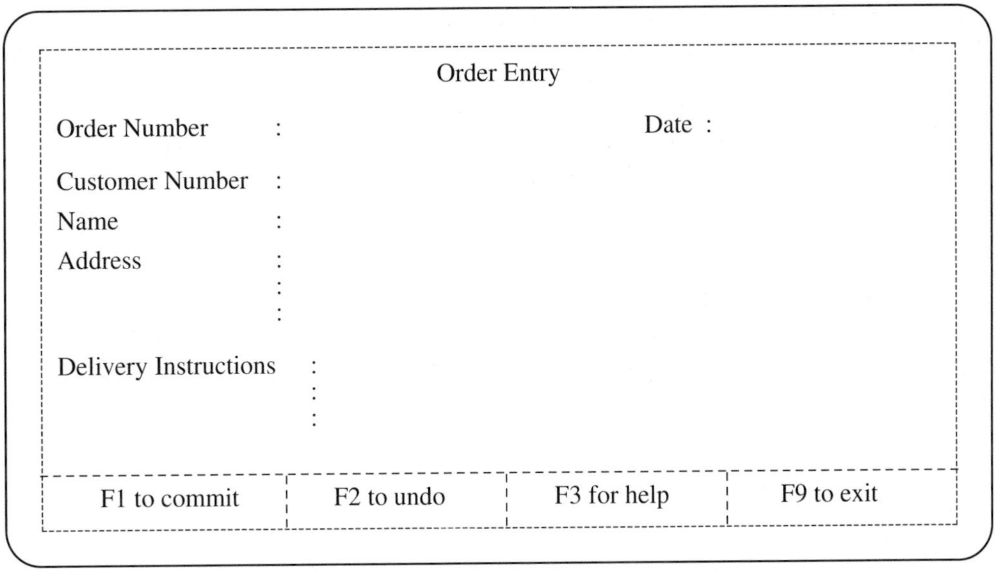

```
+--------------------------------------------------------------+
|  --------------------------------------------------------    |
|  |                    Order Entry                       |    |
|  |                                                      |    |
|  |  Order Number      :              Date  :            |    |
|  |                                                      |    |
|  |  Customer Number   :                                 |    |
|  |  Name              :                                 |    |
|  |  Address           :                                 |    |
|  |                    :                                 |    |
|  |                    :                                 |    |
|  |                                                      |    |
|  |  Delivery Instructions  :                            |    |
|  |                         :                            |    |
|  |                         :                            |    |
|  |  ------------------------------------------------    |    |
|  | F1 to commit | F2 to undo | F3 for help | F9 to exit |    |
|  ------------------------------------------------------     |
+--------------------------------------------------------------+
```

Figure 11.7   Order header entry screen

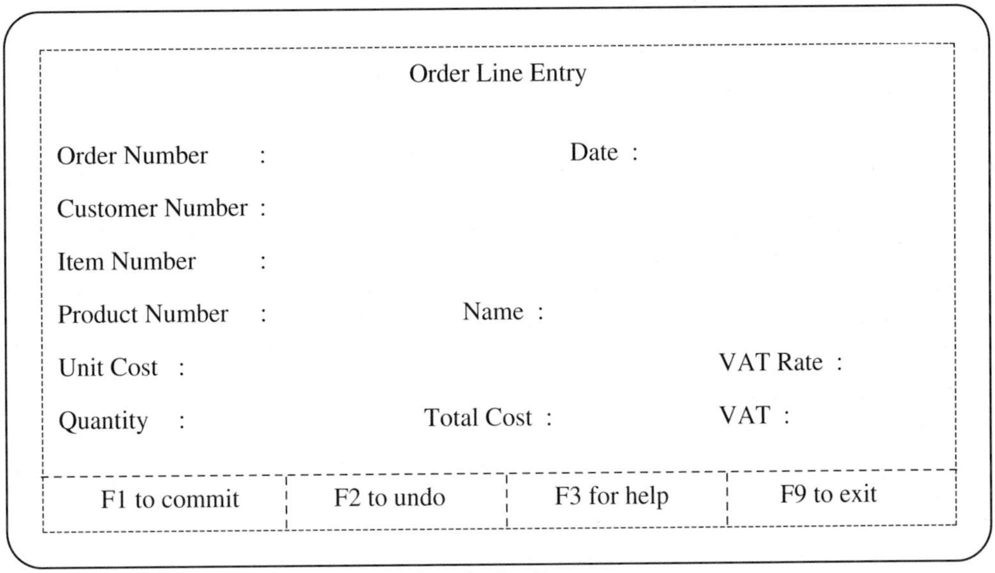

```
+--------------------------------------------------------------+
|  --------------------------------------------------------    |
|  |                  Order Line Entry                    |    |
|  |                                                      |    |
|  |  Order Number      :            Date  :              |    |
|  |                                                      |    |
|  |  Customer Number   :                                 |    |
|  |                                                      |    |
|  |  Item Number       :                                 |    |
|  |                                                      |    |
|  |  Product Number    :          Name  :                |    |
|  |                                                      |    |
|  |  Unit Cost  :                         VAT Rate  :     |    |
|  |                                                      |    |
|  |  Quantity   :        Total Cost  :     VAT  :        |    |
|  |  ------------------------------------------------    |    |
|  | F1 to commit | F2 to undo | F3 for help | F9 to exit |    |
|  ------------------------------------------------------     |
+--------------------------------------------------------------+
```

Figure 11.8   Order line entry screen

Exclusivity constraints may be difficult to implement non-procedurally. If, as in the patient database, the foreign keys for the exclusive relationships are drawn from the same domain, a single data-item, Ward-number in the entity Admission, may be used to store the Ward-number for an admission to either a maternity or gerneral ward. The corresponding dialogue need only have one field for Ward-number. Upon entry of a Ward-number, the dialogue would display the type ward as a visual check. If, however, the foreign keys are not drawn from the same domain, then two entry fields would have to be included in the dialogue which will only permit data to be entered into one of them.

11.4   COMPLETING THE SPECIFICATION

It is useful to break the more complex functions down into smaller units of processing. This identification of a function's component elements of processing offers several advantages. It:

- helps to identify lower levels of common processing;

- defines smaller units for construction and testing;

- simplifies maintenance as modifications are localised.

These advantages are most easily achieved if the structure of the function is documented effectively in some form of function model. SSADM does not specify a standard notation for this function model, although the format of the universal function model can be used to produce a specific function model. Figure 11.9 shows a sample function model for a simple payroll calculation. To be useful, the interfaces of the individual elements of the function must be carefully specified. This notation does not clearly show the invocation hierarchy of the function. Figure 11.10 shows a hierarchical module map. If JSP is being used for detailed program design, its notation provides effective documentation. Developers may, of course, choose to use their own local notation which would be

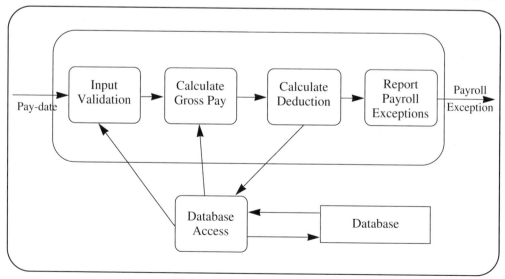

Figure 11.9   Function model

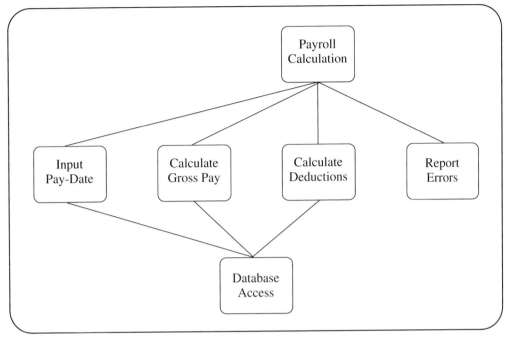

Figure 11.10   Hierarchical model map

specified in the Physical Design Strategy. It should be appreciated that the documentation requirements for functions (or elements of functions) are likely to be greater if they are implemented procedurally using a 3GL. Elements of processing which are to be implemented non-procedurally may be documented in the implementation language (eg SQL) and can thus be specified at this stage.

The earlier preparation of the Update Process Models and the Enquiry Process Models may have identified structure clashes. These have been discussed earlier in Chapter 9 and for further information the reader should consult the SSADM/3GL interface guides. The implementation environment may provide facilities which can overcome order or boundary clashes and where possible these should be used. If the implementation is with a 3GL, then interface modules need to be specified to overcome the structure clashes.

11.5   CONSOLIDATING THE PROCESS DATA INTERFACE

The Process Data Interface must be completed and verified. The target software (eg RDBMS) should be used to specify logical views based upon the implemented table types. If this is not possible or feasible, the PDI should be implemented as a set of procedural modules. As the Physical Data Design is optimised and the Physical Processes are specified, the access requirements and access paths supported by the implementation may change. It is essential that any such changes are identified and the necessary enhancements to the PDI are made. This is achieved by matching the function components specified in the FCIM against the physical data design to identify any incompatibility.

11.6   DOCUMENTATION

Physical Process Specification generates or completes several pieces of documentation:

- Application Development Standards;
- Function Definition;
- Process Data Interface;
- Requirements Catalogue;
- Function Component Implementation Map.

Three elements of the Application Development Standards are completed during Physical Process Specification, the Application Naming Standards, the Physical Design Strategy and the Physical Environment Classification. These are largely documented to local standards. The function definition is enhanced with details of any optimisation performed, the physical success units specified, common processing that has been identified and the Process Data Interface. Where necessary the Requirements Catalogue is modified to reflect any design trade-offs or compromises made. The FCIM constitutes the detailed physical process specification.

11.7   SUMMARY

As might be expected, Physical Process Specification draws upon many of the models produced earlier in life cycle. It is a procedure which is less prescriptively defined than earlier stages of SSADM and which relies heavily upon the experience of the designers and their knowledge of the development environment and the target software and hardware. The SSADM approach is to codify and document this experience and knowledge of the implementation environment in the Application Development Standards so that:

- required documentation standards are agreed for the organisation and for the project;
- a consistent approach may be used throughout;
- less experienced designers can benefit from the agreed conventions.

This is achieved by the completion of the Application Development Standards. The SSADM approach acknowledges that implementation may be procedural or non-procedural utilising sophisticated development and target environments. There are three major concerns in Physical Process Specification:

- producing program specifications from Update and/or Enquiry Process Models;
- transforming the I/O Structures into physical I/O formats;
- preparing physical dialogues from Logical Dialogue Designs.

The specification for the Process Data Interface is completed to provide a mechanism for presenting the processes with the logical structure of the data, irrespective of its physical implementation. Other issues such as common processing, error handling and system controls are addressed. The relationship between physical process and data design is important in two respects. First, optimisation should consider both the processes and the data. Secondly, care should be taken to ensure that the data structure expected by the processes is consistent with that presented by the PDI.

The relationship Physical Process Specification has with the models produced earlier in the life cycle is shown in Figure 11.11.

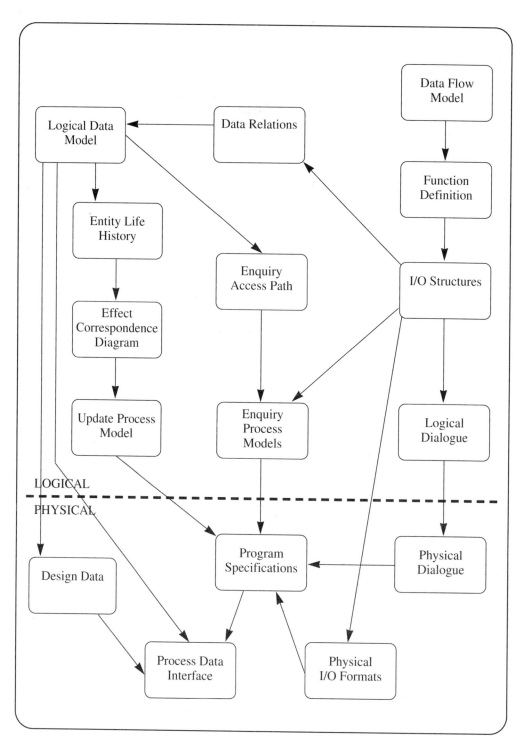

Figure 11.11   Links with earlier models

12 The Structural Model

12.1 INTRODUCTION

The five Modules of SSADM introduced in Chapter 1 represent self-contained blocks of activity:

– FS	– Feasibility Study;
– RA	– Requirements Analysis;
– RS	– Requirements Specification;
– LS	– Logical System Specification;
– PD	– Physical Design.

Within each Module there are Stages with clear inputs and outputs. These Stages are performed in a series of defined Steps, with Steps themselves decomposed into Tasks. This chapter introduces the Modules, Stages and Steps of SSADM with the objective of giving the reader an overview of the progression of systems development in SSADM, as well as indicating where the models of the previous chapters fit into the method. Detailed Stage, Step and Task definition can be found in the SSADM Reference Manual (NCC Blackwell, 1990).

Although the Feasibility Study is the first Module in the cycle, it is described last in this chapter as it effectively represents a cut-down version of the main development Stages. This relationship is clearer to see if the reader first becomes familiar with the detailed content of requirements analysis, specification and design.

The concept of the Structural Model was introduced in Chapter 1 and appropriate Structural Models are introduced in this chapter for each Stage of the development method.

12.2 MODULE RA – REQUIREMENTS ANALYSIS

This Module has five objectives:

- to determine the scope of the application;
- to establish how to integrate IT with other needs;

– to form an overall view of system costs and benefits;

– to confirm the viability of continuing further;

– to gain user ownership of the requirement.

These objectives are achieved in two Stages:

> Stage 1 Investigation of the Current Environment;
>
> Stage 2 Business System Options.

Stage 1   Investigation of Current Environment

The main objectives of Stage 1 of SSADM are to:

– describe current procedures and processes for generating, passing and storing data;

– derive a logical perspective of these procedures;

– establish information requirements not fulfilled by the current information system.

The Stage consists of five Steps:

110 Establish Analysis Framework;

120 Investigate and Define Requirements;

130 Investigate Current Processing;

140 Investigate Current Data;

150 Derive Logical View of Current Services;

160 Assemble Investigation Results.

The Products of this Stage are:

Activity Descriptions;

Activity Network;

Current Services Description;

Product Breakdown Structure;

Product Descriptions;

Product Flow Diagram;

Project and analysis scope;

Requirements Catalogue;

User Catalogue.

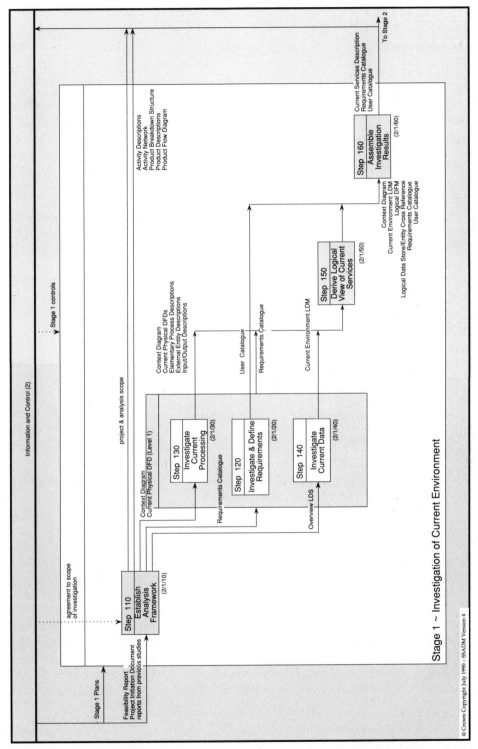

Figure 12.1   Structural Model: Stage 1 Requirements Analysis Module

The primary Techniques of this Stage are:

Data flow modelling;

Dialogue design;

Logical data modelling;

Relational data analysis;

Requirements definition.

The Structural Model for this Stage is shown in Figure 12.1.

Step 110 Establish Analysis Framework

This Step is primarily concerned with background research and familiarisation, and preparing for the detailed analysis that is to follow. A Project Initiation Document (PID) contains the terms of reference for the project and these will provide the scope, objectives, timescales, resources and constraints that will apply. A detailed project plan is prepared which schedules the activities and products determined in the project definition. These activities and products are defined in appropriate Activity Networks, Activity Descriptions, Product Flow Diagram, Product Breakdown Structure and Product Descriptions (see Chapters 1 and 13).

The primary techniques used in this step are:

– Data Flow Diagrams. A context diagram and a level 1 Current Physical Data Flow Diagram;

– A 'first-shot' overview Logical Data Structure.

These models are primarily used to confirm the system scope and boundaries specified in the PID.

The Requirements Catalogue will also begin to be compiled from information discovered in the background research and investigation.

Step 120 Investigate and Define Requirements

This Step concerns detailed examination of the current system. Principles of good system investigation must be applied here. Hence all interviews are documented and agreed by the participants and all documents, charts and files are collected and analysed (see Chapter 4 *Skidmore and Wroe*, 1988). During these interviews, two types of requirements will be established. Some will be concerned with perceived problems of the current operations, for example: the time taken to produce invoices. Others are requirements that are not fulfilled by the present system and hence represent additional services to those currently provided. All these problems and requirements are formally documented in the Requirements Catalogue. These should also be quantified where possible so that required service levels are agreed, e.g. '90% of all invoices will be raised within 2 days of order'.

The conventional fact finding and recording methods are also the basis of the Data Flow Diagrams and Logical Data Model produced in the next two Steps. The users of the system will be identified and (where appropriate) interviewed and their responsibilities documented in the User Catalogue (see Chapter 5).

Step 130 Investigate Current Processing

The level 1 Current Physical Data Flow Diagram drafted in Step 110 is now refined and decomposed using the information gained in Step 120 to produce a set of DFDs that summarise the operations of the current system. The lowest level processes on the DFDs are documented in Elementary Process Descriptions (see Chapter 3). Document Flow Diagrams can also be constructed at this step as a way of showing the information flow in the system and aiding the enhancement of the original DFD. This method of DFD development was briefly illustrated in Chapter 3.

Step 140 Investigate Current Data

A Logical Data Structure is drawn for the data in the current environment with significant attributes documented in the Data Catalogue (see Chapter 3). This LDS is an extension of the 'first shot' model of Step 110 in the light of the new information gained in the fact finding of Step 120. The data model is informally cross checked against the process model by ensuring that Elementary Process Descriptions are supported by the Logical Data Model. However, these access paths are not formally documented at this Step.

Step 150 Derive Logical View of the Current Services

This Step concerns the conversion of the Current Physical Data Flow Diagram to its logical equivalent. This Logical DFD describes the logical functions of the current system which must be carried forward to its successor. Supporting Elementary Process Descriptions and Input/Output Descriptions are amended to reflect the new diagrams. The significance and process of logical modelling is discussed in Chapters 1 and 3. An informal check is again made to ensure that the Elementary Process Descriptions are still supported by the Logical Data Model.

Step 160 Assemble Investigation Results

This Step assembles the Stage 1 products (listed below) and checks for their completeness and consistency.

The Stage 1 products are:

- Context Diagram;
- Current Environment Logical Data Model;
- Logical Data Flow Model;
- Logical Data Store/ Entity Cross Reference;
- Requirements Catalogue;
- User Catalogue.

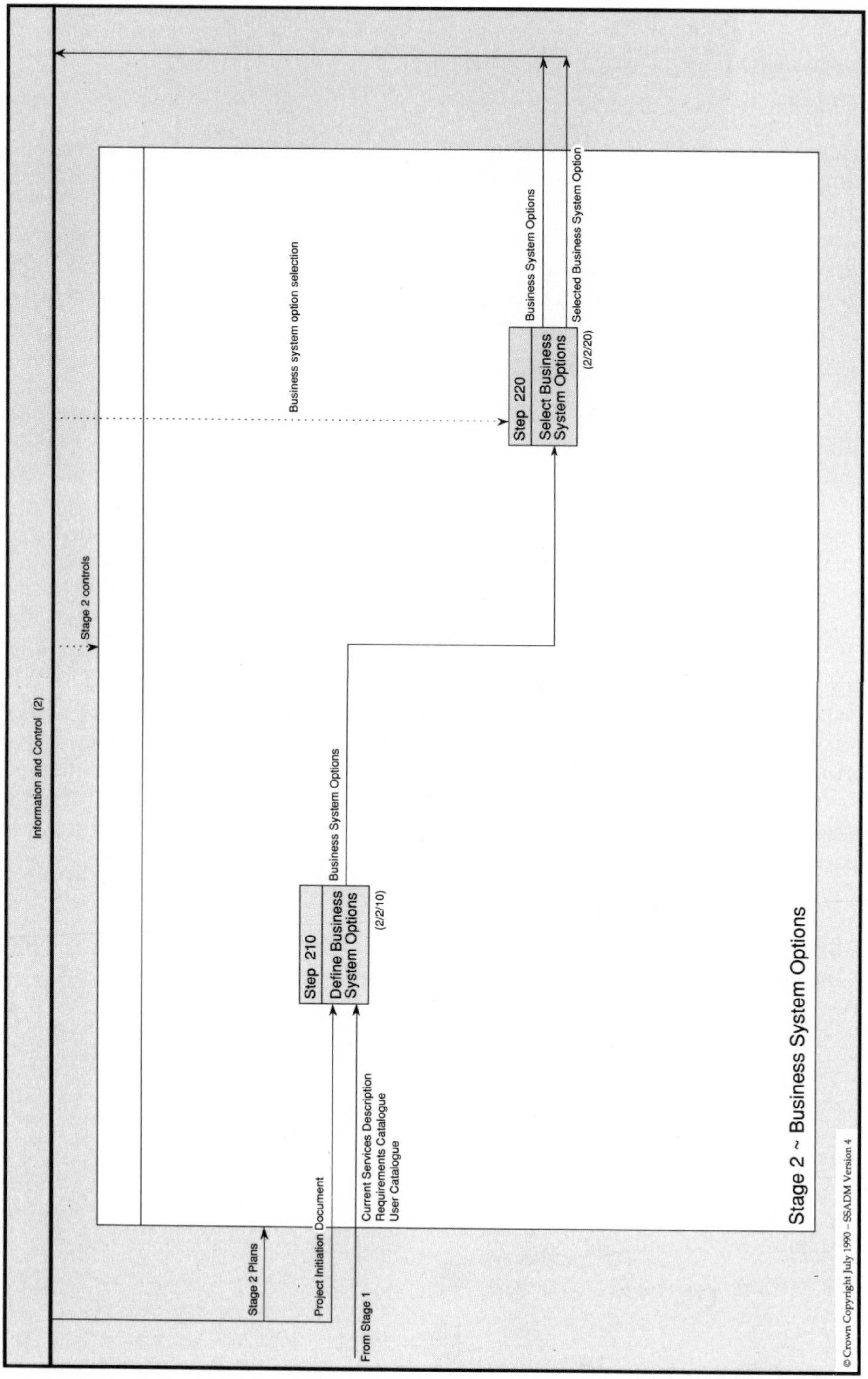

Information and Control (2)
Stage 2 controls
Business system option selection
Step 210
Define Business System Options
(2/2/10)
Business System Options
Step 220
Select Business System Options
(2/2/20)
Business System Options
Selected Business System Option
Stage 2 Plans
Project Initiation Document
From Stage 1
Current Services Description
Requirements Catalogue
User Catalogue
Stage 2 ~ Business System Options
© Crown Copyright July 1990 – SSADM Version 4

Figure 12.2 Structural Model: Stage 2 Requirements Analysis Module

Stage 2   Business System Options

The main objective of this Stage is to:

- Investigate different business system boundaries in the light of their different features, benefits, costs and risks;

- Select and justify an appropriate business system boundary on which to base subsequent information system development.

The Stage consists of two Steps:

210 Define Business System Options;

220 Select Business System Option.

The following products and techniques are used:

Products

Business System Options;

Selected Business System Option.

Techniques

Business System option;

Data flow modelling;

Logical data modelling.

The Structural Model for this Stage is shown in Figure 12.2.

Step 210 Define Business System Options

The primary objective of this Step is to develop a range of system options that meet the defined requirements. Users can then select an appropriate Business System Option (BSO) in the next Step.

It is suggested that six outline BSOs are defined that satisfy the minimum requirements defined in a prioritised Requirements Catalogue. These six options are discussed with users and two or three selected for further definition. These are described in detail with appropriate cost/benefit analysis and other organisational implications. The textual description of these selected options may be supported by boundaries superimposed upon Logical Data Models and Data Flow Diagrams to graphically show their different scope.

Step 220 Select Business System Option

This Step completes the Requirements Analysis Module. The selected Business System Option defines the boundary of the system which will be defined in the subsequent Requirements Specification Module.

12.3   MODULE RS – REQUIREMENTS SPECIFICATION

This Module has only one Stage, Specification of Requirements, hence the objectives of the Module are synonymous with those of the Stage.

Stage 3   Specification of Requirements

The main objective of this Stage is to:

- produce a complete Requirements Specification which can be used for subsequent Logical System Specification.

It is vital that the Requirements Specification has full user ownership by the time it is issued.

There are eight Steps defined in this Stage:

310 Define Required System Processing;

320 Develop Required Data Model;

330 Derive System Functions;

340 Enhance Required Data Model;

350 Develop Specification Prototypes;

360 Develop Processing Specification;

370 Confirm System Objectives;

380 Assemble Requirements Specification.

Products

Command Structures;

Menu Structures;

Prototyping Report;

Requirements Specification.

Techniques

Data flow modelling;

Dialogue Design;

Entity-event modelling;

Function definition;

Logical data modelling;

Relational data analysis;

Requirements definition;

Specification prototyping.

The Structural Model for this Stage is given in Figure 12.3.

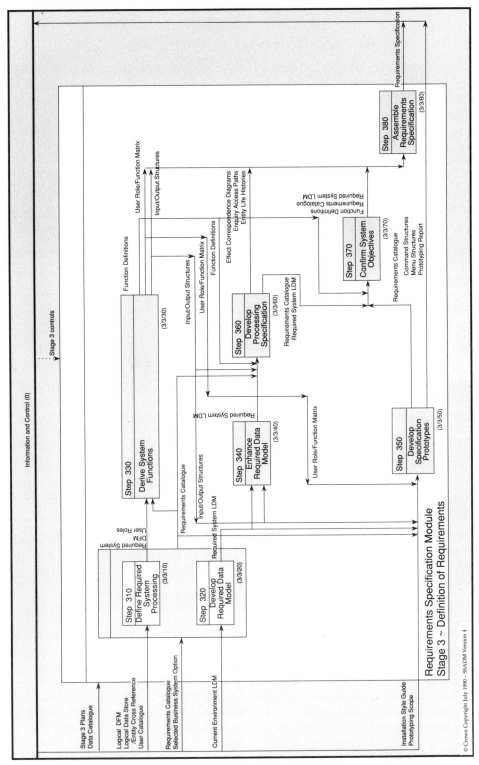

Figure 12.3 Structural Model: Stage 3 Requirements Specification Module

Step 310 Define Required System Processing

This Step amends the products of Stage 1 in the context of the Business System Option (BSO) agreed at the end of Stage 2. The Requirements Catalogue is reviewed and any requirements outside the selected BSO are noted and annotated with the reason for their exclusion.

The Data Flow Diagrams and their supporting Elementary Process Descriptions are extended/amended to reflect new processes identified within the selected BSO and to exclude processes no longer within the system boundary. The Logical Data Model is also altered to bring it into line with the agreed system definition.

Now that the system boundary is agreed the Input/Output Descriptions can be completely defined and this will demand the extension, amendment and review of previous I/O documentation.

The LDM is again cross-checked against the DFM with the formal definition of a Logical Data Store/Entity Cross Reference document.

Step 320 Develop Required Data Model

The Logical Data Model, agreed in Step 160 and amended in Step 310, is extended to support the new requirements listed in the Requirements Catalogue. Entities and Relationships can now be fully defined and relational data analysis undertaken.

An informal cross-check against Elementary Process Descriptions is again made to ensure that they are supported by the Logical Data Model. Non-functional requirements such as access restrictions and security requirements are documented in the LDM.

Step 330 Derive System Functions

The definition of functions is described in Chapter 5. This Step forms the initial construction of functions based on the Required Logical DFM and the Requirements Catalogue. Functional definition will continue through Stage 3 and is not regarded as complete until the end of Step 360.

Update functions will largely be identified from the Required System DFD and enquiries from the Requirements Catalogue or outputs on the DFD. The user interface is specified for each function using I/O Structures and the User Role/Function matrix, with critical dialogues highlighted.

Step 340 Enhance Required Data Model

The Logical Data Model is normalised. Selected input and output structures are normalised and the results integrated into the LDM.

Step 350 Develop Specification Prototypes

Prototypes have been discussed in Chapter 5. This step selects dialogues, screens, outputs menus etc, that have been deemed appropriate for prototyping. Documentation is completed before and after the prototype use/ demonstration.

Step 360 Develop Process Specification

Entity-event modelling is undertaken to provide a more detailed view of how the required system will work. Entity Life Histories (ELHs – see Chapter 7) are constructed to show how entity occurrences are created, modified and deleted. These ELHs are based on an event/entity matrix which also provides a starting point for the Effect Correspondence Diagrams (ECDs). These are produced for each event and ensure that the input data items required by the event are included in (or may be derived from) the input data items of each function that uses the event. The creation of ECDs and ELHs will feed back into the LDM and the Requirements Catalogue as the processes become better understood.

Each enquiry function will be documented in an Enquiry Access Path and the definition of these will again feedback into the LDM, where volumes are also recorded during this Step.

Step 370 Confirm System Objectives

The Requirements Catalogue is reviewed to ensure that each requirement is fully defined and satisfied in the specification of the required system. Simple cross-references are made between the Requirement Catalogue and specification products.

The Function Definitions and Required System LDM are also checked for completeness of documentation.

Step 380 Assemble Requirements Specification

This Step sees the delivery of:

- Data Catalogue;
- Event Correspondence Diagrams;
- Entity Life Histories;
- Enquiry Access Paths;
- Function Definitions;
- Input/Output Structures;
- Required System LDM;
- Requirements Catalogue;
- User Role / Function Matrix.

These products are all checked for consistency and completeness.

They form the basis of the Requirements Specification completed at the end of Stage 3.

12.4   MODULE LS – LOGICAL SYSTEM SPECIFICATION

This Module has two objectives:

- to enable management to select the technical environment which offers the best value for money in meeting their requirements;
- to provide to a physical design team an implementation independent non-procedural specification of the functionality of the proposed system.

Two Stages are defined within this Module:

Stage 4 Technical System Options;

Stage 5 Logical Design.

Stage 4   Technical System Options

The main objectives of this Stage are to:

- explore different technical products and opportunities which may be used to fulfil the Requirements Specification. These may be evaluated in terms of performance, cost, maintenance, support, skills availability, etc;

- select an appropriate technical option that will provide the hardware and software platform that will be addressed in physical system design.

Only two Steps are defined in this Stage:

410 Define Technical System Options;

420 Select Technical System Options.

The products and techniques for this Stage are as follows:

Products

Application Style Guide;

Capacity Planning Input;

Technical System Options;

Technical Environment Description (selected option).

Techniques

Dialogue Design;

Physical data design;

Physical process specification;

Technical system options.

The Structural Model for this Stage is given in Figure 12.4.

Step 410 Define Technical System Options

In this Step six outline Technical System Options (TSOs) are defined to satisfy the physical implementation of the logical Requirements Specification. These are discussed with users and two or three selected options are developed in further detail. Each TSO is described in terms of the:

- Technical environment. Type, quantity and distribution of the hardware and software;

- System description. How the system will meet the Requirements Specification;

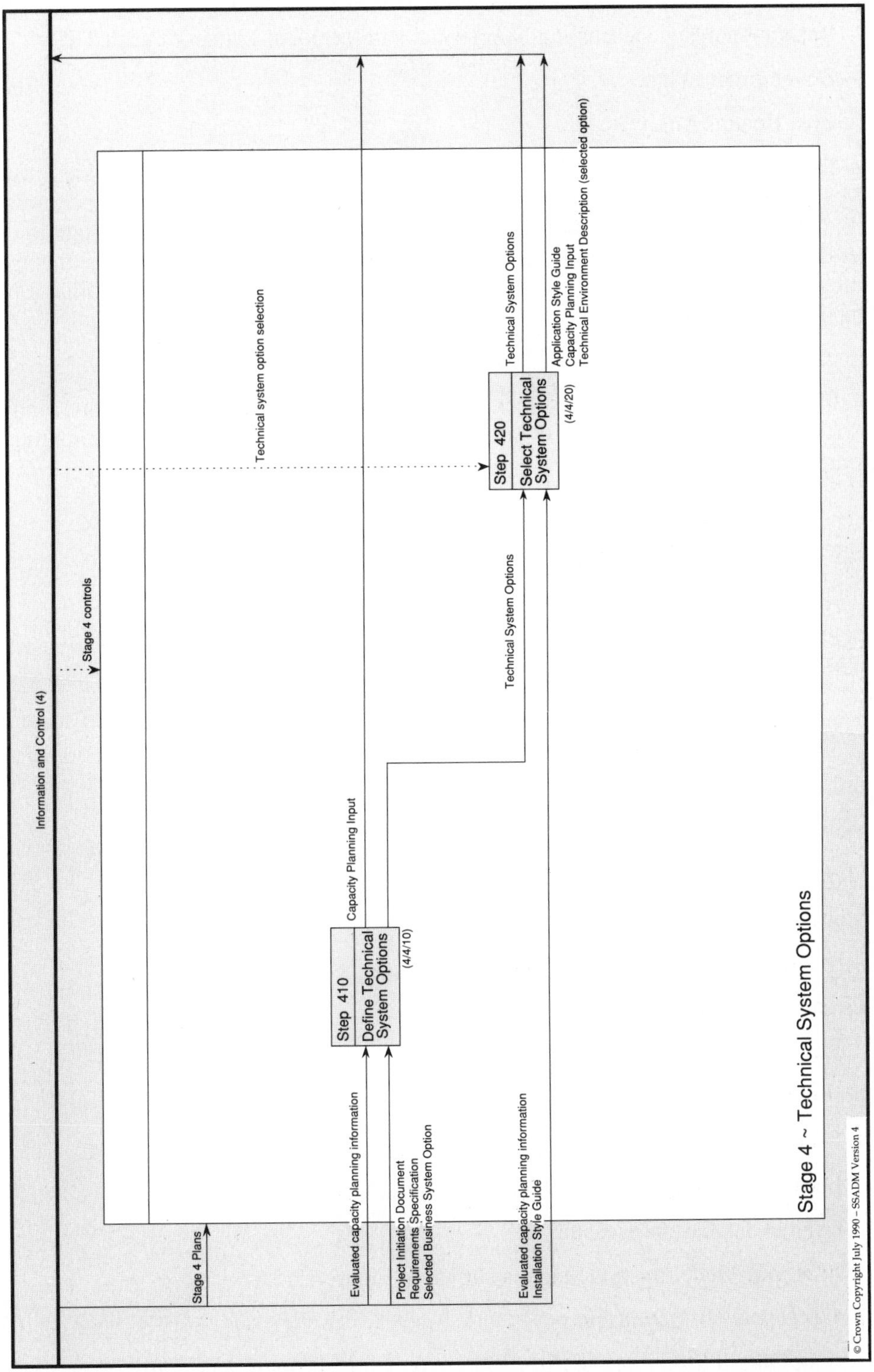
Information and Control (4)
Stage 4 controls
Stage 4 Plans
Technical system option selection
Evaluated capacity planning information
Project Initiation Document
Requirements Specification
Selected Business System Option
Step 410
Define Technical System Options
(4/4/10)
Capacity Planning Input
Technical System Options
Evaluated capacity planning information
Installation Style Guide
Step 420
Select Technical System Options
(4/4/20)
Technical System Options
Application Style Guide
Capacity Planning Input
Technical Environment Description (selected option)
Stage 4 ~ Technical System Options
© Crown Copyright July 1990 – SSADM Version 4

Figure 12.4 Structural Model: Stage 4 Logical System Specification Module

- Impact Analysis. Describing operational and technical impact of each TSO;
- Development Plan;
- Cost/Benefit Analysis.

The TSOs may be constrained by corporate strategic decisions (all hardware must be IBM and all applications implemented on the mainframe) and hence it may be unrealistic to define six TSOs. However, in other circumstances, it may be appropriate to consider different technical paths which offer genuine alternatives. For example, this is the point where a PC based Local Area Network (LAN) may be compared against a mini computer UNIX solution.

Step 420 Select Technical System Option

A Technical System Option is selected and documented. The Technical Environment Description for the selected TSO will form an important input into subsequent physical design.

Stage 5   Logical Design

The main objectives of Stage 5 are to:

- produce a logical design with detailed definition of dialogue and processing structures;
- express the design in terms which are non-procedural, independent of technical environment and give maximum opportunities for re-use.

Four Steps are used in the definition of the logical design.

Step 510   Define User Dialogues;

Step 520   Define Update Processes;

Step 530   Define Enquiry Processes;

Step 540   Assemble Logical Design.

Product

Logical Design.

Techniques

Dialogue Design;

Entity-event modelling;

Logical database processing.

The Structural Model for this Stage is shown in Figure 12.5.

Step 510 Define User Dialogues

The Input/Output Structures of Step 330 are now logically grouped into dialogue structures and elements (LGDEs – see Chapter 5) and navigation paths within each dialogue

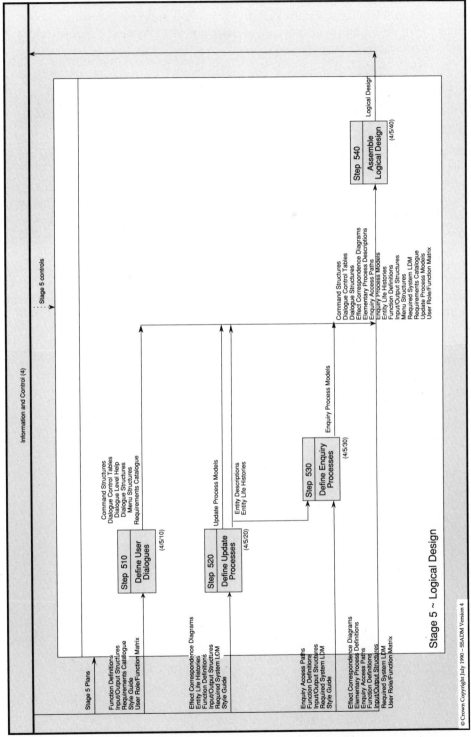

Figure 12.5   Structural Model: Stage 5 Logical System Specification Module

documented in a Dialogue Control table. Menus and Help facilities are also defined. Logical dialogue definition also occurs at this Step with physical issues of screen size, etc not introduced until Stage 6.

Step 520 Define Update Processes

In Stage 3 the required data base updates for each event are defined for each entity. At this Step the defined entity updates are consolidated into a single processing structure for each event.

State Indicators are added to the Entity Life Histories (see Chapter 7) and their meaning documented in the Entity Description. Logical Update Process Models (see Chapter 9) are constructed from the ECDs and annotated from the ELHs. At the end of this Step the specification of the data updating from each event is completed and error handling can be added.

Step 530 Define Enquiry Processes

Enquiries were also defined in Stage 3 in terms of the Input/Output Structures and Enquiry Access Paths. At this stage, a single processing structure is developed for the enquiry by combining the input and output data structures (see Chapter 9). Error handling is again defined.

Step 540 Assemble Logical Design

The products are as follows:

- Command Structures;
- Dialogue Structures;
- Dialogue Control Tables;
- Effect Correspondence Diagrams;
- Elementary Process Descriptions;
- Enquiry Access Paths;
- Enquiry Process Models;
- Entity Life Histories;
- Function Definition;
- Input/Output Structures;
- Menu Structures;
- Required System Logical Data Model;
- Requirements Catalogue;
- Update Process Models;
- User Role/Function Matrix.

These form the basis of the Logical Design completed at the end of Stage 5.

12.5   MODULE PD – PHYSICAL DESIGN

This Module only contains one Stage, Physical Design, and hence the objectives of the Module and the Stage are the same.

Stage 6   Physical Design

The main objectives of this Stage are to:

- develop a physical design that specifies the data, processes, inputs and outputs in the chosen software and hardware environment;

- produce a design that incorporates agreed installation standards;

- form a basis for system construction.

This Stage has seven defined Steps:

610   Prepare for Physical Design;

620   Create Physical Data Design;

630   Create Function Component Implementation Map;

640   Optimise Physical Data Design;

650   Complete Function Specification;

660   Consolidate Process Data Interface;

670   Assemble Physical Design.

The following products and techniques are involved:

Products

Application Development Standards;

Physical Design.

Techniques

Physical data design;

Physical process design.

The Structural Model for this Stage is given in Figure 12.6.

Step 610 Prepare for Physical Design

The main objective of this Step is to understand and document the physical environment in preparation for physical design. Facilities and constraints imposed by hardware and software will be documented and used in subsequent design stages. This documentation will be driven by:

- local installation standards;

- the formalisation of local knowledge. This often concerns performance issues in

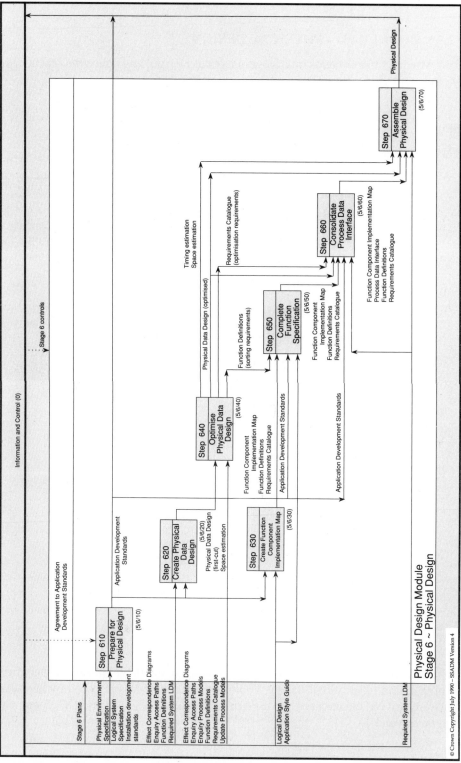

Figure 12.6   Structural Model: Stage 6 Physical Design Module

specific circumstances;

– industry guides. For example, rules for moving from Logical Data Structures to INGRES Database Management System (DBMS).

Development standards are established to cover such issues as program naming, field naming, program structuring, database naming and design, etc (See Chapter 10 and 11).

This Step will also see the initiation of user, operation and training manuals.

Step 620 Create Physical Data Design

This Step is concerned with converting the Required System Logical Data Model into a Physical Data Design using principles that are common to all DBMSs and file handlers. During this conversion, detailed requirements of physical data placement and performance are finalised leading to a data design that will reflect the specific DBMS/ file handler. Guidance on this latter conversion will have been made available in Step 610.

Step 630 Create Function Component Implementation Map

The final components of each function are defined in this Step. This will include syntax error handling, physical screen designs, input and output forms and reports. The Function Component Implementation Map (FCIM) (See Chapter 11) shows the relationship between function components and this enables the identification of common and duplicate functions. Function components that can be specified non-procedurally are defined for the physical processing system, with the exception of the database access components.

Step 640 Optimise Physical Data Design

The Physical Data Design is validated against the performance information documented in the Function Definitions and the Requirements Catalogue. Hence the process development (represented by the Function Definitions) is cross-referenced against the data development. The Physical Data Design is optimised (See Chapter 10) if the preset performance objectives are clearly not going to be achieved. This will demand time estimating and perhaps some use of capacity planning simulation.

Step 650 Complete Function Specification

This Step is only taken if the components of the Function Component Implementation Map cannot be specified non-procedurally. Program specifications are produced for the function components requiring procedural code to the level of detail required for a programmer. Outstanding structure clashes may have to be resolved.

Step 660 Consolidate Process Data Interface

The FCIM data access components are compared with the optimised Physical Data Design to identify mismatches in the 'views' of the data. Some of these mismatches will have been suggested in the Function Definition output from Step 640 as entities and relationships may have been added and removed during optimisation. This Step attempts to ensure that the functional requirements of the system are still fulfilled by the Physical Data Design.

If necessary a Process Data Interface (PDI) is defined to map the logical reads and writes of entities into physical reads and writes of database records. This PDI may be developed in an appropriate data manipulation language (such as SQL) or in a procedural model specified according to the agreed strategy of physical design.

Step 670 Assemble Physical Design

The products of the physical design Stage are checked for completeness and consistency and are then published in accordance with the standards and procedures of the organisation. The following products form the basis of this design specification:

- Function Component Implementation Map;
- Function Definitions;
- Optimised Physical Data Design;
- Required System Logical Data Model;
- Requirements Catalogue;
- Space Estimation;
- Timing Estimation;
- Process Data Interface.

This Step concludes Stage 6 of SSADM Version 4.

12.6   MODULE FS – FEASIBILITY STUDY

This Module consists of one Stage, Stage 0 – Feasibility. This Stage has four Steps:

010   Prepare for Feasibility Study;

020   Define the Problem;

030   Select Feasibility Options;

040   Assemble Feasibility Report.

A Feasibility Study is concerned with determining whether a system can be developed to meet defined business and technical objectives within the context of specified financial and operational constraints. It is recommended as a preliminary to a Full Study (Requirements Analysis, Requirements Specification and Logical System Specification) for all projects except those of low risk. Several of the models used in the Full Study (and introduced in the preceding chapters) can be employed within the Feasibility Stage.

Product

The only product of this Stage is the Feasibility Report.

Techniques

The following techniques are used:

  Business System option;

Data flow modelling;

Dialogue design;

Requirements Specification;

Technical system option.

The Structural Model for this Stage is shown in Figure 12.7.

Step 010 Prepare for the Feasibility Study

This Step determines and confirms the Terms of Reference for the project and undertakes an initial assessment of the project scope within the context of the enterprise's overall Information Systems plan. The Project Initiation Document (PID) is reviewed in preparation for more detailed analysis. An Activity Network, Activity Descriptions, Product Flow Diagram, Product Breakdown Structure and Product Descriptions (see Chapter 1) are developed for the SSADM parts of the Feasibility Study.

The assessment of project scope and complexity may be documented in the following standard SSADM products:

- Context Diagram;

- Current Physical Data Flow Diagram (Level 1);

- Overview Logical Data Structure;

- Requirements Catalogue.

Step 020 Define the problem

This Step is primarily concerned with achieving a good understanding of the business operations and information systems under consideration. The analysis of the current environment will uncover problems that must be resolved by the replacement systems as well as requirements currently unfulfilled. These problems and requirements are documented in a Requirements Catalogue. Investigation will also identify the users of current and potential systems and their responsibilities are documented in a User Catalogue. Some of the models initiated in Step 010 are extended and annotated. For example, the level 1 Data Flow Diagram may be extended to level 2 for processes that are complex, critical or unclear.

The Step culminates in a Problem Definition Statement which is a textual summary of requirements and their relative priority given business needs and objectives.

Step 030 Select Feasibility Options

This Step outlines six Business System Options (BSOs) which represent logical solutions to the requirements summarised in the Problem Definition Statement. To each of these is added a range of technical solutions and these are discussed with users. Eventually three composite options are short-listed and developed in more detail. Logical Data Structures and Data Flow Diagrams can assist the presentation of these options, graphically showing differences in scope and facilities.

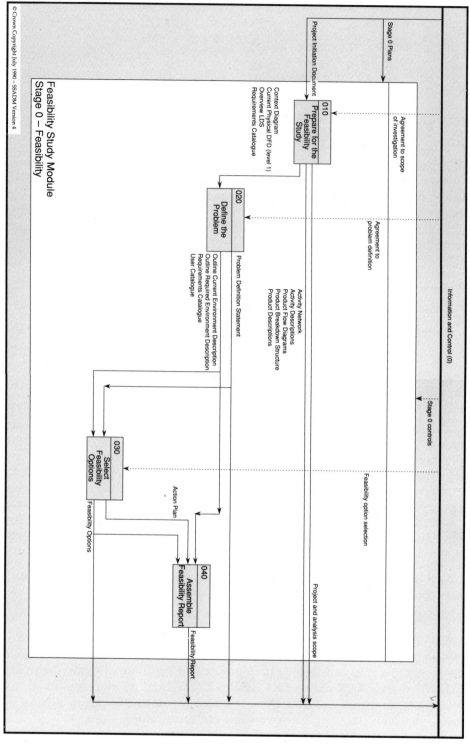

Figure 12.7   Structural Model: Stage 0 Feasibility Study Module

Each short-listed option includes an appraisal of its financial and operational effects as well as its relationship to the Project Initiation Document and the enterprise's overall Information Systems strategy.

Step 040 Assemble Feasibility Report

The Feasibility Study Module is completed by this Step. The products of the Stage are reviewed for completeness and consistency. These products (listed below) will form the basis of a published Feasibility Report:

- Action Plan;
- Feasibility Options;
- Outline Current Environment Description;
- Problem Definition Statement;
- Requirements Catalogue;
- User Catalogue.

12.7 SUMMARY

This chapter has briefly introduced the SSADM Structural Model. It serves as a reminder of the techniques used in this book as well as placing them in their development framework.

13   Project and Quality Management

13.1   INTRODUCTION

This chapter examines selected issues of project management and quality control. It is not designed to be a detailed examination of either of these areas, but rather a summary of the project and quality controls that will surround the products described in the rest of this book.

A project may be defined as a set of activities to develop a set of products using appropriate resources within an agreed budget. Thus a project management method must be concerned with Activities, Organisation, Products, Plans, Controls and Quality Assurance.

13.2   ACTIVITIES

The structure of SSADM Version 4 is defined in terms of the products, or deliverables, required. From this comes an analysis of the activities needed to produced the products.

There are two types of activities. The first type are those required to satisfy the user's requirements to a defined quality. The second type are those required to manage the project. The former are defined within the Structural Models of SSADM and the latter are defined by the Project Management methodology adopted for the project.

In SSADM the delivery of the Module products, to the agreed quality standards, marks the completion of that Module. Each Module comprises one or more Stages; each Stage has a single goal and encompasses a set of Steps which are directed toward a particular activity. An activity is achieved through a set of tasks which transform input products into output products.

The Structural Model (see Chapter 12) covers all the SSADM Modules and Stages. It shows how products are derived from each other through a logical series of transformations. Each transformation acts upon one set of products to create another set of products. An Activity Network is developed for the project to define the sequence and the dependencies of technical and quality activities.

13.3   ORGANISATION

The project should be organised to make the most effective use of the technical skills and abilities of staff assigned to the project. Everyone associated with the project must be clear about their responsibilities to ensure that management of the project is both efficient and responsive.

It is recommended that a Project Board is appointed to oversee the entire project development. This committee will be composed of management, user and technical representatives. Its main responsibility is to ensure that technical and business integrity is maintained for each Module, and to sign off products and give authorisation to proceed with the project. This Project Board must have sufficient authority to commit resources or else its usefulness is reduced.

A Project Manager is appointed by the Project Board with responsibility for the day-to-day running of the project as well as for overall planning and control.

The Technical Team is responsible to the Project Manager for developing the products for each Module. The skills required in this team will vary from Module to Module and with the size and nature of the project. The composition of the team may change from Module to Module.

For large or complex projects, where the technical skills required will differ according to the Module under development, a Module Manager, reporting to the Project Manager, may be appointed to take responsibility for the successful completion of a Module. In this case the Team Leaders will report to the Module Manager.

It is also suggested that there are significant benefits from allocating teams by functional area rather than by tasks. However, this will normally require one team to have the added role of co-ordinating the products of the other teams to ensure consistency across tasks.

Two supporting activities are also required. These are Quality Assurance and Project Administration.

The Quality Assurance team is responsible for monitoring the integrity of the project. Like the Project Board it requires business, user and technical members who should stay with the project throughout its life to ensure continuity. The business representatives ensure adherence to schedules and budgets, the users guard their interests and requirement, and the technical representatives monitor technical competence. Further issues of Quality Assurance are discussed later in this chapter.

The Project Support Office should provide a number of administrative functions or skills. It may supply staff for the Quality Assurance team as well as providing expertise in estimating, planning and control for a number of projects. In other circumstances, it may only have a restricted role providing the secretarial and configuration management activities for an individual project. Configuration management is concerned with the procedures for controlling the products of a project and for providing information about the status of the products as they are developed.

However these support activities are organised, access to them must be defined by the Project Board at the initiation of a project. Figure 13.1 shows the structure of the project organisation.

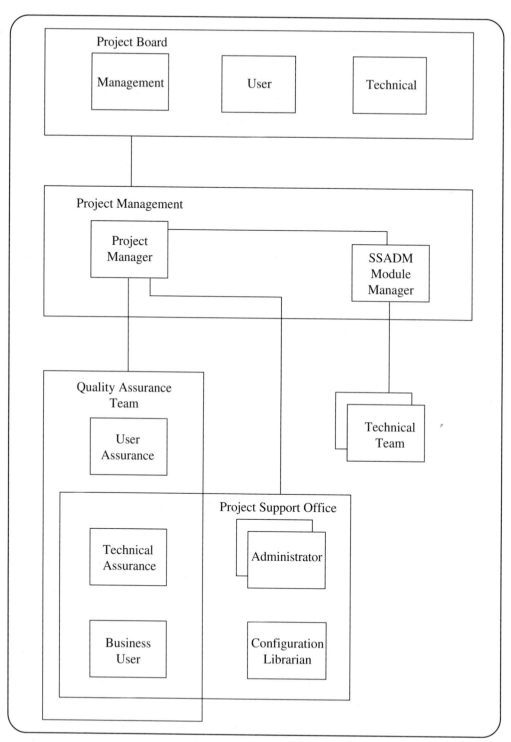

Figure 13.1 Project organisation structure

13.4   PRODUCTS

Products fall into three categories:

- – those required to facilitate effective management of the project;
- – those required to fulfil the technical requirement of the project itself;
- – those that show that quality has been built into the system.

The hierarchical arrangement of these products can be shown in a Product Breakdown Structure (PBS), the top level of which is shown in Figure 13.2. For each project the PBS defined within the SSADM Dictionary should be reviewed with the aim of including any additional project specific products.

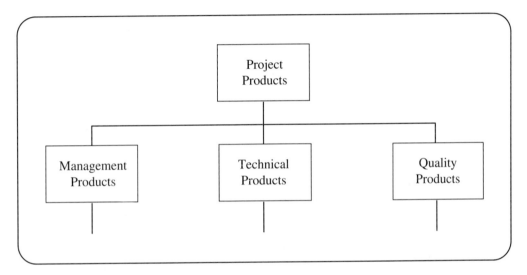

Figure 13.2   Project products

The product breakdown for the management products is shown in Figure 13.3. The project is instigated by a Project Initiation Document (PID) accompanied by a record of the Installation Development Standards for specification and implementation of applications. The PID defines the terms of reference for the project. It will also include the business systems needs, objectives for the project, risks, organisation and responsibilities and any constraints within which the project must progress.

It is essential that accurate records of decisions made at Project Board meetings are maintained to aid effective project management. Checkpoint Reports and planned Highlights Reports should also be prepared, providing the Project Board with a summary of the current project status.

On completion of the project the Project Evaluation Review is compiled, using progress reports as a basis. This provides an assessment of the management procedures used during the system development. Documenting this information should enable experience learnt on the project to be used to improve procedures for future projects.

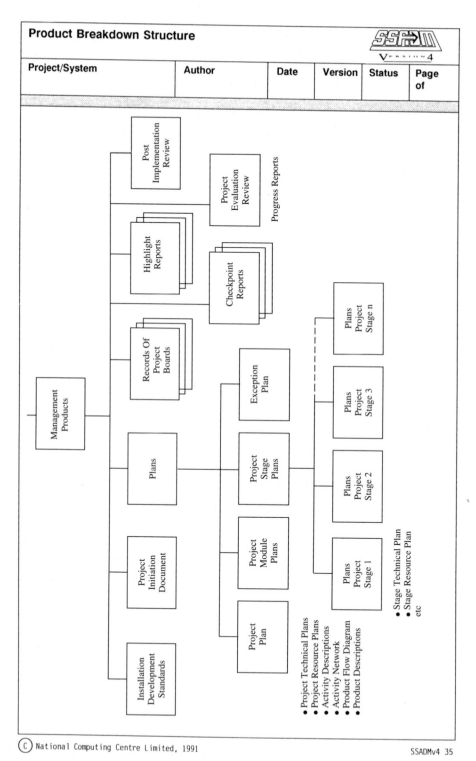

Figure 13.3   Management Products

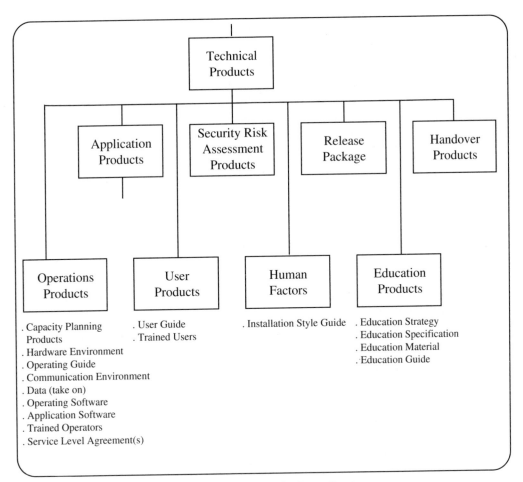

Figure 13.4   Technical products

The Post Implementation Review consists of the documents which are produced during the project and are used to show the products, activities and resources required to manage and maintain the system on completion of the project. It is used to control the subsequent stages of the system's life.

The Technical Products are those produced to meet the business needs as defined by the objectives of the project. The top level PBS, Figure 13.4, shows the major products of this development process. The Application Products are those associated with the development of a computer system using SSADM.

13.5   PLANS

It is important to plan how to meet identified targets in terms of products, timescales, costs and quality. The process of planning involves estimating, collating, sequencing, scheduling and assigning the project's resources to deliver the desired products.

Associated with planning is the definition of tolerances by specifying the limits for resources and timescales within which the project or Modules of the project must operate. Tolerances may be defined as percentages of budget or time.

Plans are prepared at three levels of detail: Project, Module and Stage. The Project Plan, prepared by the Project Manager, shows the major technical activities and the resources required for the whole project. It provides a fixed reference point against which progress can be measured and requires approval by the Project Board.

Before the start of each Module the Project Manager prepares a Module Plan for approval by the Project Board. If necessary, a Stage Plan is prepared by the Project and Module Managers which covers specific activities within an SSADM Stage.

The lowest level Plan should provide a sufficiently detailed activity breakdown so that activities can be directed and controlled effectively. Each Plan is reviewed against the higher level Plan and any inconsistencies reconciled.

For each level of Planning two types of Plan are identified:

- a Technical Plan concerned with the products to be delivered and with the activities necessary to ensure that products emerge on time to the required quality;

- a Resource Plan concerned with finance, resource allocation and the effective use of skills needed to carry out the Technical Plan.

A Technical Plan will be accompanied by appropriate supporting documents such as:

- Product Breakdown Structure;

- Product Descriptions;

- Activity Network;

- Activity Descriptions.

In addition, a supporting narrative may be required justifying the plans, with details of responsibilities, control tolerances, the quality control strategy and methods of monitoring and reporting performance.

Initially the plans document the projected resource used over time. As the project progresses these are updated with actual use. If the updates identify that the permitted tolerance levels have been, or are about to be, exceeded, then an Exception Plan may be raised. This will describe the circumstances and propose corrective action in the form of a new Module or Stage Plan which is submitted to the Project Board for approval.

The product oriented nature of SSADM has been stressed throughout this book and so Project and Module Plans are, not surprisingly, product driven. A Product Breakdown Structure is derived for each product and a Product Flow Diagram can be produced showing the dependencies of product development. Once these dependencies between products and product components are derived then Activity Networks can be defined. The form of an Activity Network will follow local standards. As soon as activities are in a logical sequence then timescales can be estimated, work scheduled and resources allocated.

Estimating the resources required and the duration of a system development is discussed elsewhere (see References). However, it must be recognised that structured methodologies change the balance of effort required in the various phases of development as well as permitting a more formal basis for estimation. In general, more time is given to the earlier phases of the project producing a detailed logical design. This should be compensated for by an expected reduction in the time spent in the physical design and construction stages. Furthermore, detailed estimating in these later stages should become more accurate because consistent methods of specification will permit the development of reliable metrics. However, knowledge gained from previous experience demands that the criteria used in their construction must be recorded. This provides information to assess any variance of the actual from the estimate so that the estimating procedure can be refined by the Project Support Office.

A bottom-up approach may be applied by aggregating the estimates of all the activities needed to produce the required products. These estimates can then be converted to time and resource requirements and accumulated to produce estimates for the individual Steps, Stages and Modules, and finally for the project as a whole. Initial estimates, required for the project level plans, are likely to have a wide margin of error and this should be recognised in the plan. As the project progresses and more detailed information becomes available, then estimates can be refined to reduce the error margins.

Activity descriptions		Duration (staff-days)
Interview staff-internal	A	24
Investigate lease position	B	16
Interview staff-external	C	20
Write initial audit report	D	8
Survey computer contracts	E	28
Produce standard contract	F	24
Document staff interviews	G	20
Produce standard job descriptions	H	12
Write final report	I	8

Figure 13.5   Activity descriptions

Once the effort for each activity has been estimated, it is possible to use these in the Activity Network and determine the duration of the project with the given resources. Figure 13.5 shows the activity descriptions for an example project and Figure 13.6 illustrates their inter-relationship in an Activity Network.

The estimated duration of this project is 84 staff-days.

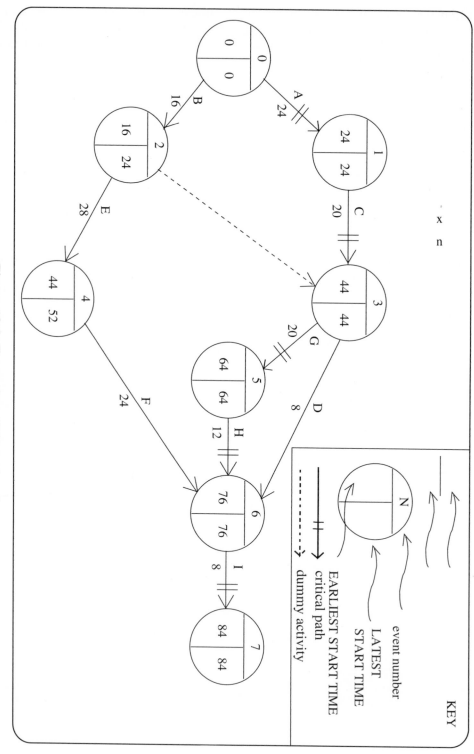

Figure 13.6   Example activity network

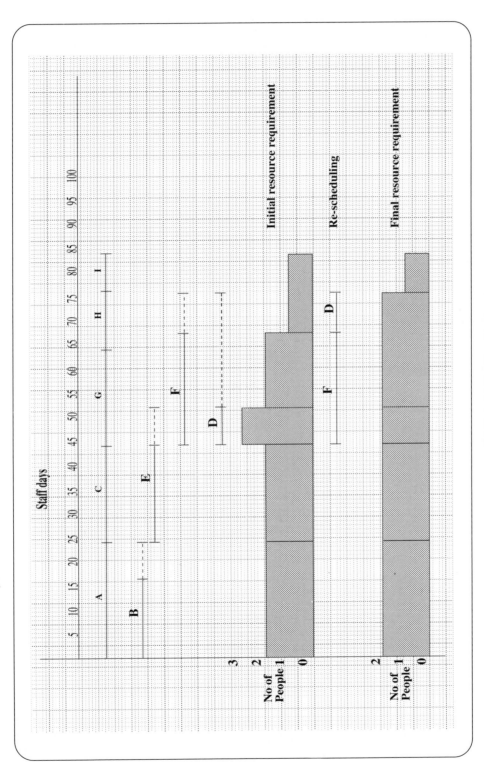

Figure 13.7 Gantt chart

In other circumstances, the required duration of the project can be specified and the resources required calculated so that the specified project duration is not exceeded.

A Gantt chart can be used for resource smoothing, that is adjusting the scheduling of activities, within the limits of their float, so that fluctuations in individual resource requirements are reduced. This is illustrated in Figure 13.7 where the resource requirement is reduced from three to two staff by re-scheduling activity D.

The application of cost factors to the required effort makes it possible to calculate the probable cost of the project. However, project management overheads are not included as elements in the estimating guidelines and these should be added to the costing. It should be remembered that quality assurance tasks require a significant resource allocation, not only from the project team but from quality assurance administration and users.

Planning is not a serial process; it is normally necessary to iterate between the various processes of estimating and scheduling in order to achieve a satisfactory initial plan with agreed completion dates and costs. As more experience is gained during the stages of the project these initial estimates may be re-evaluated and modified.

Once the effort and other resources required have been estimated a project cost can be derived. Each organisation will have standard costing algorithms which may be applied. A graph of costs against time, with the Module boundaries identified, should be prepared as the Resource Plan.

13.6   CONTROLS

One of the most obvious measures of productivity is the completion time of the project. However, the use of a structured method need not necessarily lead to reduced development times. Other benefits may be gained such as increased user satisfaction, greater reliability and easier and cheaper maintenance – in short, better systems quality. These benefits are more likely to be realised if the project is effectively monitored and controlled.

The method of project management selected at the beginning of the project will define the type of management control and the frequency and formality of reporting. For example, Mid-Module and End of Module reporting may be required by the Project Board, whereas the Project Manager will require (perhaps less formal) regular weekly or fortnightly checkpoints. The final review of the project marks the project closure and, if satisfactory, the delivery of the system to the user. At this final review the date and arrangements for the Post Implementation review will be settled.

To monitor the project effectively, it is important to anticipate possible deviations from the project plan. This requires that when the current status of a product is reported the effort needed to complete the product (and hence the Stage and Module) is also identified.

This information can be plotted on the Technical Plan bar chart and an assessment made of the project status. The project management may then re-evaluate the Stage or Module Plans and submit an Exception Plan to apply any necessary corrective action. Corrective action may include switching resources between tasks, applying more resources, reviewing the estimates and replanning.

A project may also have product and configuration management controls. Product controls are concerned with quality reviews and technical exceptions and both aim to identify and correct errors as quickly as possible.

Quality reviews examine a product to ensure that it is complete and correct and meets the defined quality criteria. The conduct of these reviews is discussed later in this chapter.

Technical exceptions are concerned with unplanned circumstances relating to one or more products. This may be when a product or product component fails to meet its specification or where there is a request for change.

Change, and how those changes are controlled, is one of the major problems encountered with larger projects. Significant changes must be sanctioned by the Project Board although less significant amendments may be agreed at a lower level as long as the project will remain within the tolerance levels allowed. However, care must be taken that the overall progress of the project is not impeded by continual change requests. All changes must be clearly specified and costed in the light of their impact on the project products and business area.

If the cumulative effect of acting on a technical exception will cause the plan tolerance to be exceeded, then an Exception Plan must be prepared and submitted to the Project Board for a decision.

Configuration management controls the development of products by providing a formal mechanism for labelling products, their development status, and the relationship between them. Every product which forms part of the delivered system is a configuration item. These products are held in the configuration library and managed by a librarian.

13.7   PROJECT MANAGEMENT: SUMMARY

A project exists to deliver specific products to fulfil a business need. Moreover, a project will have to progress within certain constraints in terms of time, cost and resources. In order to manage a project it is necessary to have:

- an organisational structure with defined roles and responsibilities;
- a set of activities which can be monitored and controlled;
- plans against which progress of product delivery can be assessed.

The benefits of employing a structured project management method should include:

- completion on time and to budget;
- production of quality products;
- separation of management and technical aspects of organisation, planning and control;
- control over the work progress and authorisation to proceed or not;
- timely and appropriate involvement of senior management;

- timely and appropriate involvement of users;
- visibility of the project's progress;
- definition of agreed communication channels.

Many UK government projects have now adopted PRINCE (Projects in Controlled Environments), the origins of which are in the PROMPT (Project Resource Organisation Management and Planning Techniques) as a methodology for project control. PRINCE offers a framework within which projects can be correctly specified, designed and implemented.

13.8   QUALITY ASSURANCE

It is increasingly important to ensure that computer systems and their associated business procedures satisfy agreed quality criteria.

Quality is concerned with two main issues.

Functionality. From the user's perspective the system should achieve predefined objectives in terms of content, reliability and performance.

Construction. Fitness for purpose is important but software quality involves other criteria. For example, it may be necessary for the design to display agreed standards of the following qualities:

- economical;
- documented;
- testable;
- comprehensible;
- reusable;
- flexible;
- resilient;
- easy to use;
- modular;
- correct;
- maintainable;
- reliable;
- portable;
- efficient.

These criteria may not be of equal importance in all projects. If it is intended that the delivered system will only operate on one hardware and software platform, then system portability is irrelevant. However, these criteria should ensure that user's requirement are not met at great cost to the overall soundness of the software.

Quality Assurance (QA) is concerned with ensuring that the end product conforms to agreed standards and requirements. Quality Control (QC) comprises those QA actions that enable the control and measurement of agreed quality criteria.

Good QA ensures that the system is constructed according to agreed standards, easing the software development process and subsequent maintenance. The application system will, no doubt, be modified to maintain its usefulness as user's requirements change (adaptive maintenance) or to correct errors or omissions (corrective maintenance). It is thus important that the application system should be capable of accommodating these new requirements with the minimum of difficulty.

13.9   QUALITY AND SSADM

The quality of delivered products is viewed as critically important in SSADM. Each SSADM product has quality criteria defined in the Product Descriptions and an example for Logical Data Structures is shown below (see Figure 13.8).

Is the variant identifier completed correctly?
Is each entity genuinely a thing of significance about which information needs to be held?
Are all entity names singular and meaningful?
Does each entity have a unique identifier?
Is each relationship genuinely a significant association between entities?
Is each relationship end named and capable of being read accurately and sensibly?
Does each relationship begin at an entity and end at an entity?
Do all relationship ends in an exclusive arc have the same optionality?
Have all 1:1 relationships been resolved?
Have all m:n relationships been resolved?
If mandatory relationships, will there always be an occurrence of the entity at the other end?
Are any relationships redundant?
Is the structure consistent with the previous version of the LDS?
Are all the entities in Third Normal Form?

Figure 13.8   Quality Criteria for Logical Data Structures

Those criteria that relate to the syntax of the model may be enforced by using CASE tools. Other criteria which relate to the way the LDS has captured the semantics of the application are only effectively checked by peer and user review.

Three types of product are developed when using SSADM:

- a deliverable which will be utilised by a subsequent Module;

- a deliverable which will be used by a subsequent SSADM Step;

- a product which is developed as a by-product of a technique or procedure. For example, the event-entity matrix used in the construction of Entity Life Histories.

The first two must be subject to formal quality control procedures whilst informal review may be suffiicient for the last category.

13.10   PLANNING FOR QUALITY

A clearly defined approach to QA is required if quality is to be maintained. This may be achieved by establishing Quality Assurance Departments or Groups. These should consist of personnel experienced in the use of the techniques and methodology. It may be appropriate to have one or two permanent QA staff with additional staff being seconded to the department as necessary. This rotation of membership helps to dispel the 'us and them' syndrome. To avoid conflicts of interest, QA staff should not be members of the project team whose project is being examined.

Quality assurance should be considered at two levels. The activities developing the products must fulfil quality standards and the products produced must reach specified levels of quality.

Ensuring that a product is delivered with the necessary quality involves certain steps:

- that quality criteria for the product are agreed at the beginning of the project;

- that these quality criteria must be allocated priority;

- that steps are taken to ensure these criteria are met during the production of the product as well as achieved by the product itself;

- that it is possible to demonstrate that the product meets these criteria.

For a quality criterion to be useful its attainment must be measurable. The issue of metrication is complicated by the subjective nature of many of the criteria associated with analysis, design and construction. However, where possible appropriate metrics should be specified clearly indicating the required level of performance.

Aspects of a system's performance such as response time are relatively easily quantified and the ability of a product to satisfy such criteria can be clearly demonstrated. Others, such as usability, are less easy to measure meaningfully and uniformly. The development of appropriate metrics, their application and refinement justifies the existence of a dedicated metrics group. De Marco's adage 'You can't control what you can't measure' reinforces the importance of metrics in systems development.

Quality is more likely to be achieved if the developers are familiar with the quality criteria and standards that have to be met and appreciate the importance of Quality Assurance. The development process becomes less frustrating and more satisfying in an environment geared to producing quality products.

13.11 METHODS OF QUALITY CONTROL

The methods of Quality Control used on the project will be dictated by the Quality Assurance approach adopted and are likely to include both testing and reviews.

13.11.1 Testing

Some products, particularly those developed in the later stages of the life cycle, require some form of testing to check whether quality criteria have been met. Software testing involves ensuring that certain test data is processed correctly and may be supported by software tools (for regression testing perhaps). When tests have been satisfactorily completed, the tested product should still be formally reviewed with the test data as one of the review inputs.

13.11.2 Review

The most common approach is to formally review or walkthrough products at the end of given Stages or Steps in the project. The general structure of such reviews will be specified in a quality standard. However, the allocation of roles to project members, the composition of the review team and the timing of the reviews will vary.

QA procedures may be particularly difficult to specify for small projects. If every product is reviewed at length the QA overheads may exceed the resources available. It is especially important in these circumstances to keep review teams to a minimum size and to concentrate on key products. This is not to suggest that the quality of some products is unimportant but rather that they should be dealt with by informal walkthroughs. An informal review has the same objectives as a formal review but will not have the same resource and administrative overheads.

The informal walkthrough is, in fact, useful on projects of all sizes. Products should be reviewed informally at intermediate stages to maintain the levels of quality throughout the project. A great deal can be gained by walking through a product with one or two project members. Such informal walkthroughs should also ease the passage of products through the subsequent formal reviews.

The benefit of QA is not just to deliver the system the user wants but also to trap problems, errors or omissions as early as possible during the project. This minimises the cost of correcting these problems. It is generally accepted that the earlier errors are detected, the lower the cost of correction. The role of the review in error detection must be appreciated. Tom de Marco references Jones' study of formal code inspection. This examination of around 10 million lines of installed code concluded that inspection can remove as many as 85% of all defects. Although it is recognised that such inspections are time-consuming they "virtually always pay back more than they cost" (de Marco – see References). Although these results are concerned with program code inspection, it should be evident that formal reviews of analysis products will lead to even greater gains in quality and productivity. Consequently, this chapter is concluded by a summary of the roles required for review inspections.

13.12   QUALITY REVIEW ROLES

There are three main roles associated with a quality review – chairman, presenter and reviewers.

The Chairman

The chairman should be a member of the project team and is responsible for management and running of the review. The chairman requires both tact and authority to ensure that the review does not degenerate into a self-congratulatory chorus or the verbal assassination of the presenter or one of the reviewers. The chairman need not be the most senior member present but must have both technical and managerial skills.

The Presenter

The presenter will normally be the author of the product. Presenting a product for review is a difficult task. The presenter should be aware of the standards that have to be achieved and must endeavour to ensure that the product meets these prior to the review. It is the responsibility of the project manager to ensure that the team members have been given necessary training in presentation.

The Reviewer

The reviewers are required to help improve the product. Before the meeting they should be supplied with review checklists and details of the quality criteria that the product should meet. They should not use the review as an opportunity to publicise their own solutions but rather to help detect errors. In some instances alternative solutions may be suggested. These should be noted by the chairman and should not be allowed to dominate the meeting.

Reviewers are selected because they are experienced in the preparation of the product under review or they are familiar with the project or because they represent an alternative viewpoint. QA department staff are useful review panel members. Users should also be actively involved in formal reviews. They are likely to contribute the most detailed and relevant understanding of the application area.

It can be difficult to give precise guidelines for the selection of reviewers but a commonly agreed convention is the exclusion of the presenter's line manager. If the presenter feels that the review may be used to determine promotion or salary then the objectives of the review will be subverted. This can be avoided by not inviting the presenter's manager or other individuals who have a direct influence over the career of the presenter.

13.13   PLANNING AND RUNNING REVIEWS

Each review or walkthrough should follow a pre-defined procedure. There are three steps for each review: preparation, review meeting and follow-up. Each of the roles has specific responsibilities at each stage.

13.13.1   Preparation

The review should be scheduled so that the presenter has time to prepare the presentation material and the reviewers are able to read and comment on all the preliminary documentation.

The chairman should:

– check with the presenter that the review item is ready for review;

– set an agreed preparation time with the presenter (and reviewers if necessary);

– issue review invitations, review checklists and copies of the product;

– produce an agenda in consultation with the presenter;

– give the presenter an opportunity to rehearse the review if necessary. This is particularly important if the presenter is unfamiliar with this role.

The presenter should:

– liaise with the chairman to agree a preparation period, date and agenda for the review;

– prepare and rehearse the presentation. Presentation standards should be specified by the QA department;

– consider any comments made by the reviewers prior to review.

The reviewers should:

– familiarise themselves with the quality criteria and the product;

– check the product for completeness, errors, omissions and deviations from the standards, recording issues on an error list;

– give the presenter a copy of the error list prior to the review;

– send the chairman the error list and annotated product if unable to attend.

13.13.2   The Review Meeting

The review meeting should last no more than two hours. When meetings last longer they become unproductive.

The chairman should:

– Clearly state the objectives at the beginning of the meeting;

- Decide whether sufficient reviewers are present to conduct the review satisfactorily and if to postpone the review;
- Ask the presenter to give an overview of the product only allowing interruptions for clarification;
- Ask the reviewers for general comments which should be clarified by the presenter;
- Decide whether the review should continue or whether the presenter should be asked to rework the product for subsequent re-review;
- Ask the presenter to give a detailed walkthrough of the product discussing issues as they arise;
- Ensure that the reviewers have an opportunity to raise all the necessary issues;
- Ensure that all errors or problems with the product are carefully documented together with the action required for their correction;
- Decide whether the product should be re-reviewed either because the corrections to the errors require checking or because the review has not been completed;
- Agree with the reviewers the responsibility for signing off all errors noted.

The presenter should:

- present a brief overview of the product describing why it was produced, the approach taken and any assumptions made;
- note comments made by members of the review team;
- walkthrough the product in detail answering any points raised by the reviewers;
- collect from the chairman details of any follow-up actions, error lists and annotated copies of the product.

The reviewer should:

- raise any general concerns or issues at the beginning of the meeting;
- raise specific issues during the walkthrough;
- not assess the presenter or the style in which the product is documented unless it deviates from agreed standards.

13.13.3   Follow-up

The objective of the follow-up is to correct any errors highlighted during the review.

Chairman

The Chairman should normally sign off the required actions that were specified at the review and when all these have been agreed the product can be signed off. If some errors cannot be resolved in the follow-up period, then these should be reported to the project manager who may have to allocate more time for the completion of this product.

Presenter

The Presenter must correct the errors found during the review so that the product can be signed off.

Reviewer

On occasions a reviewer may have been nominated to sign off errors found during the review.

13.14   DOCUMENTING QA

There are several types of documentation for QA. First, the quality criteria and standards have to be clearly defined. This is the province of the QA department or group and is one aspect of setting up a quality assurance system. These criteria and standards are relatively static and are applicable to all projects. Secondly, project specific QA requirements are established at the beginning of the project with the advice of the QA department. This may include the modification of standard review checklists or the addition of certain procedures that reflect that particular environment. During the project each review or walkthrough will produce several documents. These include the signed-off documented review product, the error lists and the follow-up action list. These documents permits the audit of project quality.

13.15   QUALITY ASSURANCE: SUMMARY

The quality of an application system or one of its intermediate products is measured in terms of its ability to satisfy specified requirements. Primarily, the system must satisfy the user's requirements in terms of functionality, information handling, performance, storage requirements and security considerations. However, other quality criteria are concerned with build quality, adaptability and longevity. These must underpin the functional objectives of the software.

Quality Assurance addresses the issue of ensuring that the necessary levels of quality are met. This may be achieved by a quality department or group which will specify quality criteria and the standards that have to be achieved. Walkthroughs of individual products or groups of products are an effective mechanism for checking quality. These reviews may be formal or informal and their timing and scope will vary from project to project. However, their effectiveness must not be underestimated.

The roles of the participants in a walkthrough are clearly defined, together with their responsibilities before, during and after the review. These procedures should be documented in a quality standard and staff given appropriate training so that they can fulfil these roles.

Within SSADM, quality criteria are defined for the SSADM products and form the basis for Quality Assurance procedures. These criteria are contained within the Product Descriptions.

14 Review

14.1 INTRODUCTION

This book concentrates upon the modelling techniques central to SSADM Version 4 and relates them to the SSADM Structural Model. References have been made to the Version 3 approach where this is notably different. In this last chapter, we review the main changes and features of the modelling methods and philosophy of SSADM Version 4.

First, it is worth examining the overall objectives usually shared by systems developers. These include:

- to deliver the system to users on time and within budget;
- to deliver systems that meet users' agreed needs;
- to deliver systems able to change to reflect alterations in the business environment;
- to improve the effective and economic use of available skills;
- to improve system quality by reducing specification errors;
- to manage and control the development process;
- to provide comprehensive and accurate documentation;
- to avoid lock-in to a single source of supply.

The effectiveness of a particular systems development methodology might be measured by the degree to which it achieves these objectives. However, there is no agreed or reliable way of measuring the performance of a methodology against these or other criteria. In fact, the market in systems development methodologies exists, in part at least, because no approach can unequivocally be shown to be the best or most correct. Rather, it is generally agreed that a methodological approach should be used and that this should be developed in the light of experience.

Most methodologies, including SSADM Version 4, would claim to achieve the first seven objectives to some degree. However, the last objective is best achieved by a non-proprietary methodology such as SSADM.

14.2   REVIEW OF MODELS

14.2.1   Data Flow Diagrams

The Data Flow Diagram is traditionally one of the central analysis and design models of systems development, focusing on the process or functional perspective of the system. In some methodologies, it remains a central model throughout the whole of the life cycle. In principle Data Flow Diagrams are easy to develop and interpret. However, in practice individual developers will construct them at different levels of detail and this can cause difficulties in quality control. Furthermore, trainee analysts are often unsure of the level required of them. This variation is addressed in SSADM by defining a Function which is independent of the diagram construction and this is illustrated in Chapter 5.

In SSADM Version 4, the importance of the Data Flow Diagram is reduced as the life cycle progresses and in the later stages its role is taken over completely by Function Definitions and I/O Structures. This reflects the perception that the DFD is a powerful analysis and end-user tool but that its usefulness as a detailed design model is limited.

Within SSADM Version 4, the notation has been further clarified and extended to increase its descriptive power and the guidelines for preparing Logical DFDs from their physical equivalents have been improved.

14.2.2   Logical Data Modelling and Relational Data Analysis

A complementary data modelling perspective is provided in virtually all methodologies by some variant of entity relationship diagramming supported by normalised data structures.

In SSADM Version 3, Logical Data Modelling was treated as a separate activity to normalisation (Relational Data Analysis – RDA). This resulted in two data models which were combined to form a third model, the Composite Logical Data Design (CLDD). Many practitioners found this approach artificial and cumbersome. In Version 4 the principles of RDA have not changed but its application is now earlier in the life cycle. This gives a more natural approach to data modelling in that RDA is now an integral part of the preparation of the Logical Data Model. Furthermore, the definition of the relationships can be derived more confidently and correctly.

In Version 3 ELHs defined from the Logical Data Structure (LDS) had to be reviewed and redrawn following the construction of the CLDD. This post-CLDD review added a significant overhead to the development of Entity Life Histories. This does not apply any more.

The notation used in Logical Data Modelling has been enriched to capture more existence and integrity issues. Furthermore, the naming of the relationships has been standardised to allow more effective communication. Early verification of the Logical Data Model is now formalised by the development of Enquiry Access Paths.

14.2.3   Function Definition

The central role of the Function Definition as a 'super-technique' has no real equivalent in SSADM Version 3. It groups together the user's functional requirements described in different SSADM product perspectives. In this way, it acts as a cross-reference document between events, I/O structures, on-line dialogues and DFD process specifications. The point has already been made that the Function Definition effectively overcomes the problem of Data Flow Diagraming at different levels of detail.

14.2.4   I/O Structures and Dialogues

Dialogue design has traditionally been a difficult modelling area. The notation of the Logical Dialogue Outline was not clearly defined or developed in Version 3 and hence was not widely used. The Jackson based notation used in Version 4 is certainly more graphical but could perhaps be extended to increase its descriptive power. We have included (see Chapter 5) a simple extension using the Structure Diagram notation to specify error handling. Across methodologies, there is a notable lack of a commonly accepted model for dialogue design. It will be interesting in the future to evaluate the use and success of this new model in SSADM.

14.2.5   Entity-Event Modelling

Entity-event Modelling is fundamental to SSADM Version 4 providing a coherent progression from analysis to design. The two components of entity-event modelling are Entity Life Histories (ELHs) and Effect Correspondence Diagrams (ECDs). Although ELHs were defined in Version 3 they were not always used in practice. Furthermore, even when they were developed on projects the post-CLDD review was rarely fully completed.

Entity Life Histories were often complicated in Version 3 because they were usually applied to unnormalised data. Hence the technique was found to be difficult to learn and apply. In SSADM Version 4, ELHs are developed for a normalised data model and experience suggests that these will produce relatively simple ELHs. In the examples in this book (see Chapter 7) we had to introduce derived and control fields in order to illustrate more complex aspects of the notation and construction

Furthermore, ELHs were often perceived as an end in themselves. In SSADM Version 4 the ELH is used as a basis for process design and so their importance is more evident to the developer.

The second component of entity-event modelling is the Effect Correspondence Diagram. This is a new technique which validates the ELHs and hence gives the practitioner more confidence in their construction. Effect Correspondence Diagrams (ECDs) are subsequently developed into Logical Update Process Models.

14.2.6   Logical Database Process Design

The use of the Structure Diagram in Logical Database Process Design is a significant departure from Version 3, which used the form-based approach provided by Logical Process Outlines (LUPOs and LEPOs). The graphical approach of Version 4 is more suitable for CASE support and could be used as a basis for automatic code generation. The natural progression of Logical Process Design from the earlier models (ECDs, I/O Structures and EAPs) provides a clear path for the definition of the processes. Unfortunately, this also means that if any of the earlier models are not completed then process design will have to be approached differently. On occasions developers may find a textual description of the process model useful and we have suggested the Action Diagram as an appropriate format.

14.2.7   Physical Design

The emphasis on the refinement of Application Development Standards demonstrates the flexible approach to Physical Design in SSADM Version 4. In particular, the Physical Environment Classification and the Physical Design Strategy formally summarise product specific knowledge and help standardise the local approach to Physical Design. This includes the definition of the requirements for the Process Data Interface which can be used to maintain a high degree of correspondence between logical and physical designs. The Installation Style Guide helps to codify the installation's approach to Human-Computer Interaction and should ensure that interface design is consistent.

14.3   OTHER ISSUES

The change from the task oriented approach to systems development of Version 3 to the product oriented perspective of Version 4 should provide a better opportunity for the successful integration of Project and Quality Management. Project progression is tied to deliverables and this requires an approach which properly selects and sequences the activities required to produce those deliverables, as well as defining standards for their acceptance.

This is a change not just for the systems developer but also for Information Systems and User management. The specification of quality criteria in the Product Descriptions and the concept of the Information Highway contribute more explicitly defined control and reporting procedures.

A strength of Version 4 is the way models produced early in the life cycle are used to develop subsequent models. Any automated tool must support and exploit this characteristic. Furthermore, computer support is essential to eliminate repetitive redrawing and validation of diagrams.

SSADM Version 4 is more relevant than Version 3 to Fourth Generation Language (4GL) implementation and particularly acknowledges the need to accommodate product specific features in the design approach. However, the explicit consideration of some of the issues in process design, such as Structure Clashes, are more appropriate to Third Generation Languages (3GLs), such as COBOL, than for Fourth Generation Languages.

Prototyping has been recognised as an important approach to systems development for some time. In Version 4, prototyping guides and documentation is provided although currently its use is only emphasised in dialogue design. Incremental development using prototyping could be accommodated via the Physical Design Strategy.

14.4   CONCLUSION

SSADM Version 4 is intended to be a core methodology which is supplemented further by SSADM Guides. Guides can be developed by any interested party, but the Design Authority Board policy is to recommend Guides which actively reinforce the Method. Guides are being prepared which may be used to integrate other methods or techniques or increase the life cycle coverage, for example, Strategic Analysis. Interface Guides will define the relationship between SSADM products and proprietary software.

SSADM Version 4 effectively builds upon the strengths of Version 3. It continues to emphasise the importance of integrating the process, data and event perspectives of a system and encourages maximum user involvement. The increased integration of techniques reduces the discontinuity between logical and physical design and highlights the need for automated support. Version 4 is less dogmatic than its predecessor, reflecting both the way the methods are actually used by developers as well as recognising differences in project scope, environment and application.

References

GENERAL

Ashworth C, Goodland M, *SSADM: A Practical Approach*, McGraw Hill, 1990

Ashworth C, (Aims Syst., London U.K.), 'A more polished method [SSADM]', *Syst. Int. (UK)*, Vol. 17, no. 2, pp. 37–40, February 1989

Edwards H M, Thompson J B, Smith P, (Sch. of Comput. Studies & Math., Sunderland Polytechnic U.K.), 'Experiences in the use of SSADM: series of case studies 1/ First-time users,' *Information Software Technology* (UK) Vol. 31, no. 8, pp. 411–419, October 1989

Edwards H M, Thompson J B, Smith P, (Sch. of Comput. Studies & Math., Sunderland Polytechnic U.K.), 'Experiences in the use of SSADM: series of case studies 1. Experienced users', *Information Software Technology* (UK) Vol. 31, no. 8, pp. 420–428, October 1989

Longworth G, Nicholls D, *SSADM Manual (Version 3)*, Vol. 1 Tasks and Terms, NCC, 1986

Longworth G, Nicholls D, *SSADM Manual (Version 3)*, Vol. 2 Techniques and Documentation, NCC, 1986

CCTA, PRINCE (Projects in Controlled Environments), Vols 1–5, NCC Blackwell, 1990

CCTA, SSADM Version 4 Reference Manuals, Vols 1, 2, 3 & 4, NCC Blackwell, 1990

DATA FLOW MODELLING

De Marco T, *Structured Analysis and System Specification*, Prentice-Hall 1979.

Gane C, Sarson T, *Structured Systems Analysis*, Improved System Technologies, 1980

McMenamin S, Palmer J, *Essential Systems Analysis*, Yourdon Press, 1984

Ward P, Mellor S, *Structured Development for Real-Time Systems*, Vol. 1 Introduction & Tools, Yourdon Computing Press, 1985

DATA MODELLING AND PHYSICAL DATA DESIGN

Benyon D, *Information and Data Modelling*, Blackwell Scientific, 1990

Chen P, 'The Entity-relationship Model – toward a unifying view of data', *ACM Transactions on Data Base Systems*, Vol. 1, no. 1, March 1976, pp 9–36

Chen P, E-R, 'A historical perspective and future directions', in *Entity-Relationship Approach to Software Engineering*, C G Davis et al Eds, North Holland, 1983.

Codd E F, 'A relational model of large shared data banks', *Communications of ACM*, Vol. 13, no. 6, June 1970, pp 377–387

Codd E F, 'Further normalisation of the data base relational model', *Courant Computer Science Symposium 6: Data Base Systems*, Prentice-Hall, May 1971 pp 65–98

Date C J, *An Introduction to Database Systems*, Vol I, 5th edition, Addison Wesley, 1991.

Everest G, *Database Management Objectives, System Functions, and Administration*, McGraw-Hill, 1986

Fagin R, 'Multivalued dependencies and a new normal form for relational databases', *ACM Transactions on Database Systems*, Vol 2, No 3, Sept 1977, pp 262–278

Fagin R, *Normal Forms and Relational Database Operators*, Proceedings of 1979 ACM SIGMOD International Conference on Management of Data.

Howe D R, *Data Analysis for Data Base Design*, 2nd edition, Edward Arnold 1989

Kent W, 'A simple guide to five normal forms in relational database theory', *Communications of ACM*, Vol 26, No 2, Feb 1983, pp 120–125

Korth H, Silberschatz A, *Database Systems Concepts*, McGraw-Hill, 1986

Martin J, McClure C, *Diagramming Techniques for Analysts and Programmers*, Prentice-Hall, 1985

Weiderhold G, *Database Design*, 2nd edition, McGraw-Hill, 1978

FUNCTIONS, DIALOGUES AND PROTOTYPING

Alavi M, 'An assessment of the prototyping approach to information systems development', *Communications of the ACM*, Vol. 27, No. 6, June 1984

Boar B, *Application Prototyping*, Wiley-Interscience, 1984

Coates & Vlaeminke, *Man-Computer Interfaces: An introduction to Software Design and Implementation*, Blackwell, 1987

Lantz K E, The *Prototyping Methodology*, Prentice-Hall, 1986

Martin J, McClure C, *Diagramming Techniques for Analysts and Programmers*, Prentice-Hall, 1985

Skidmore S, 'Extending the LDO notation', Assist Partnership, *Research Paper 2*

STRUCTURE DIAGRAMS, ENTITY-EVENT MODELLING AND PROCESS DESIGN

Burgess R, *Structured program design using JSP*, Hutchinson, 1987

Cameron J R, *JSP and JSD: The Jackson Approach to Software Development*, 2nd edition, IEEE Computer Society Press, 1989

Jackson M A, *Systems Development*, Prentice-Hall, 1983

Jackson M A, *Principles of Program Design*, Academic Press, 1975

Martin J, McClure C, *Diagramming Techniques for Analysts and Programmers*, Prentice-Hall, 1985

Skidmore S, Wroe B, *Introducing Systems Analysis*, NCC Blackwell 1988

PHYSICAL PROCESS SPECIFICATION

Martin J, *Fourth Generation Languages*, Vol. 1, Prentice-Hall, 1985

Martin J, Leben J, *Fourth Generation Languages*, Vol. 2, Prentice-Hall, 1986

Martin J, McClure C, *Action Diagrams, Clearly Structured Specifications, Programs and Procedures*, 2nd edition, Prentice-Hall, 1989

Page-Jones M, *The Practical Guide to Structured Systems Design*, 2nd edition, Prentice-Hall International Ed., 1988

Skidmore S, Wroe B, *Introducing Systems Design*, NCC Blackwell, 1990

PROJECT AND QUALITY MANAGEMENT

De Marco T, *Controlling Software Projects, Management, Measurement and Estimation*, Yourdon Press, 1982

Freedman D P, Weinberg G M, *Handbook of Walkthroughs, Inspections and Technical Reviews*, 3rd edition, Little, Brown and Company, 1982

Gilb T, *Principles of Software Engineering Management*, Addison-Wesley, 1988

Yourdon E, *Structured Walkthroughs*, 2nd edition, Yourdon Press, 1978

Index